SpringerBriefs in Cognitive Computation

About the Series

SpringerBriefs in Cognitive Computation are an exciting new series of slim high-quality publications of cutting-edge research and practical applications covering the whole spectrum of multi-disciplinary fields encompassed by the emerging discipline of Cognitive Computation. The Series aims to bridge the existing gap between life sciences, social sciences, engineering, physical and mathematical sciences, and humanities.

The broad scope of Cognitive Computation covers basic and applied work involving bio-inspired computational, theoretical, experimental and integrative accounts of all aspects of natural and artificial cognitive systems, including: perception, action, attention, learning and memory, decision making, language processing, communication, reasoning, problem solving, and consciousness.

More information about this series at http://www.springer.com/series/10374

Andrew Abel · Amir Hussain

Cognitively Inspired Audiovisual Speech Filtering

Towards an Intelligent, Fuzzy Based, Multimodal, Two-Stage Speech Enhancement System

Springer

Andrew Abel
Department of Computing Science
and Mathematics
University of Stirling
Stirling
European Union

Amir Hussain
Department of Computing Science
and Mathematics
University of Stirling
Stirling
European Union

ISSN 2212-6023 ISSN 2212-6031 (electronic)
SpringerBriefs in Cognitive Computation
ISBN 978-3-319-13508-3 ISBN 978-3-319-13509-0 (eBook)
DOI 10.1007/978-3-319-13509-0

Library of Congress Control Number: 2015946085

Springer Cham Heidelberg New York Dordrecht London

Printed on acid-free paper

Springer International Publishing AG Switzerland is part of Springer Science+Business Media (www.springer.com)

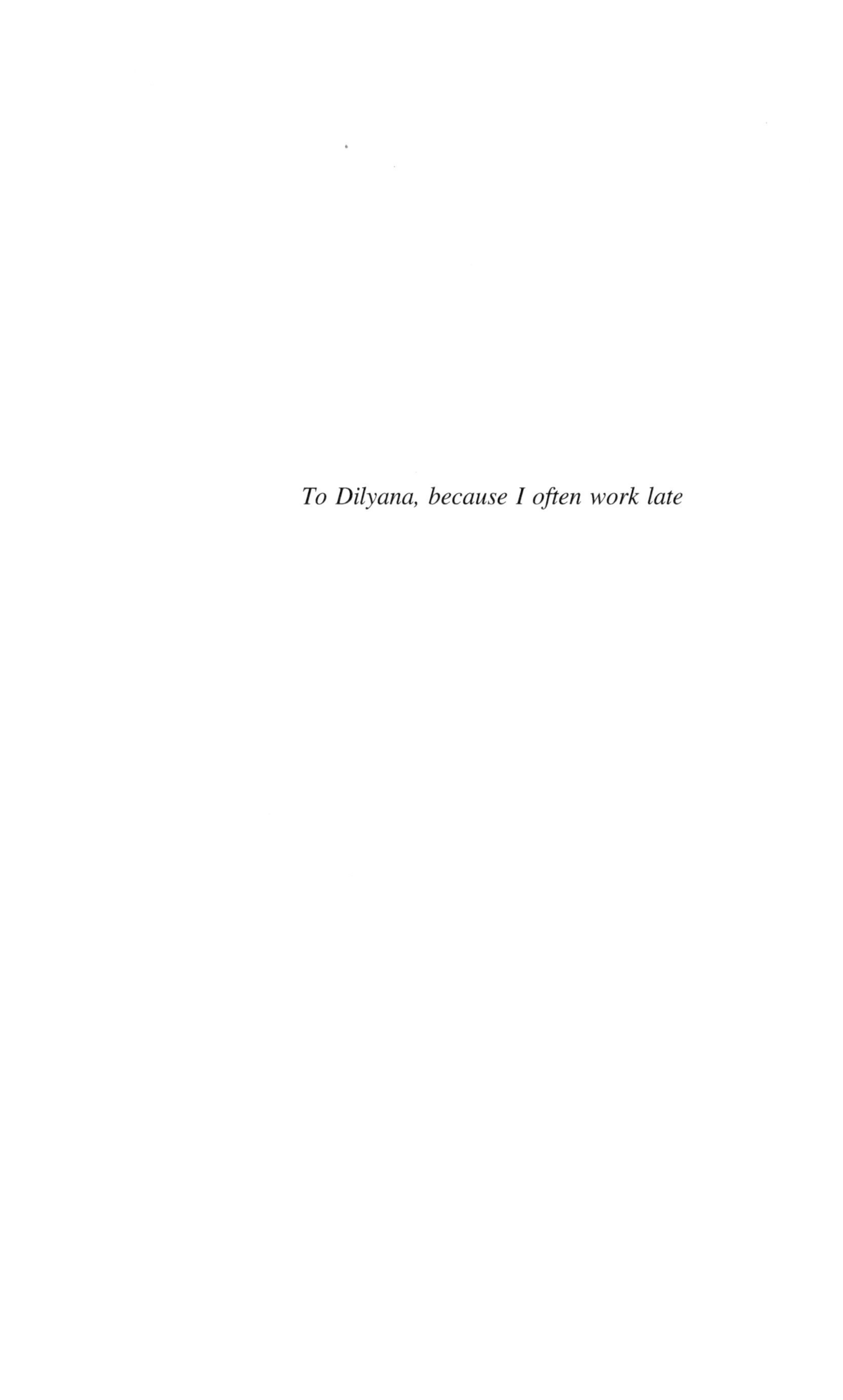

To Dilyana, because I often work late

Foreword

This is an excellent text book that combines theory and practice, suitable for Master and Ph.D. students as well as senior researchers. It fulfils a scientific community necessity.

Matar
September 2014

Marcos Faundez-Zanuy

Preface

This book focuses on the area of cognitively inspired multimodal speech processing, and is the result of several years of research by the authors, which focuses on many disparate areas, such as cognitive inspiration, speech processing, image processing, and machine learning.

This book presents a novel two stage multimodal speech enhancement system, making use of both visual and audio information to filter speech, and explores the extension of this system with the use of fuzzy logic to demonstrate proof of concept for an envisaged cognitively inspired autonomous, adaptive, and context-aware multimodal system. The concept of single modality two stage filtering is extended to include the visual modality. Noisy speech information received by a microphone array is first pre-processed by visually derived Wiener filtering. This pre-processed speech is then enhanced further by audio only beamforming. This results in a system which is designed to function in challenging noisy speech environments (using speech sentences with different speakers from the GRID corpus and a range of noise recordings), and both objective and subjective test results show that this initial system is capable of delivering very encouraging results with regard to filtering speech mixtures in difficult reverberant speech environments. Some limitations of this initial framework are identified, and the extension of this multimodal system is explored, with the development of a fuzzy logic-based framework and a proof of concept demonstration implemented. Results show that this proposed autonomous, adaptive, and context-aware multimodal framework is capable of delivering very positive results in difficult noisy speech environments, with cognitively inspired use of audio and visual information, depending on environmental conditions.

This book is aimed at providing a comprehensive introduction to the field of cognitively inspired audiovisual speech processing. As there are many different facets of this field, including audio-only speech processing, image tracking, ROI extraction, and fuzzy logic, there are very few examples of research where all of these aspects are combined together to present a single comprehensive reference. This book therefore contains fully referenced and easily accessible summaries of all of these areas, along with an introduction into the cognitively inspired basis behind multimodal speech filtering, and combines them with research into a cognitively inspired speech processing system. We also include guidance to performing objective and subjective speech evaluations. It is hoped that those that are interested in this fascinating area can use the guides and the research presented in this book as the basis of their own research.

Stirling, Scotland
September 2014

Andrew Abel
Amir Hussain

Acknowledgments

Over the years I was preparing this book, so many people have helped me that I couldn't possibly begin to name them all, so if I've missed anyone out, I'm sorry! First of all, I have to thank my Ph.D. supervisor and co-author Amir Hussain, who gave me this opportunity in the first place, and who has continued to provide me with so many other opportunities! I want to thank Leslie Smith, whose tips for surviving a conference I faithfully (and occasionally nearly fatally) followed throughout my career so far, and who I've continued to develop my research with. Oh and Kate too, who was always ready to help whenever I turned up at her office with a spreadsheet and a look of abject terror.

I couldn't have followed this line of work without the ESF Cost 2102 events that I was involved with, and I'm particularly grateful for the guidance from Anna Esposito and Marcos Faundez-Zanuy. I also have to thank my friends who I made, who reminded me that we really were all in it together, David, Marco, Alina, Erik, and Maria (and her lovely neck!), to name just a few, and who made the (ahem) ordeal of having to fly to a European city twice a year and spend a week working very hard, going to bed early, and doing a tiny, tiny bit of moderate socialising, into some of the highlights of my Ph.D.!

I'd like to thank my friends for their looks of disbelief and heavy sighs, and my family for their occasional warm support, and frequent sarcasm. Probably most of all though, if it wasn't for the support of Laura, I would never have made it this far. I really don't know where I'd be without our late night text conversations and her patience, sympathy, and good advice, and I wish her nothing but happiness.

—Andrew Abel

Thanks to my ever persevering family.

—Amir Hussain

Contents

Acronyms

2D-DCT	Two Dimensional Discrete Cosine Transform
AAM	Adaptive Appearance Models
AcAMs	Active Appearance Models
ACM	Active Contour Models
ANC	Adaptive Noise Canceller
ANNs	Artificial Neural Networks
ASA	Auditory Scene Analysis
ASMs	Active Shape Models
ASR	Automatic Speech Recognition
AVCDCN	Audio-Visual Codebook Dependent Cepstral Normalisation
AVICAR	Audio-Visual Speech Recognition in a Car
BANCA	Biometric Access Control for Networked and E-Commerce Applications
BM	Blocking Matrix
BSS	Blind Source Separation
CCA	Canonical Correlation Analysis
CDCN	Codebook Dependent Cepstral Normalisation
CRM	Coordinate Response Measure
CUAVE	Clemson University Audio Visual Experiments
DCT	Discrete Cosine Transform
DFT	Discrete Fourier Transform
EM	Expectation Maximisation
FBF	Fixed Beamformer
FIR	Finite Impulse Response
FPGAs	Field-Programmable Gate Arrays
GMM	Gaussian Mixture Model
GMM-GMR	Gaussian Mixture Model—Gaussian Mixture Regression
GMMs	Gaussian Mixture Models
GMR	Gaussian Mixture Regression
GSC	Generalised Sidelobe Canceller
HMM	Hidden Markov Model

HMMs	Hidden Markov Models
ICA	Independent Component Analysis
IIFastICA	Intelligently Initialised Fast Independent Component Analysis
IS	Itakura-Saito Distance
ITU-T	International Telecoms Union
LIF	Leaky Integrate-and-Fire
LLR	Log-Likelihood Ratio
LPC	Linear Predictive Coding
LSP	Line-Spectral Pairs
M2VTS	Multi Modal Verification for Teleservices and Security Applications
MAP	Maximum a Priori
MCMC-PF	Markov Chain Monte Carlo Particle Filter
MFCC	Mel Frequency Cepstral Coefficients
MLP	Multi Layer Perceptrons
MLR	Multiple Linear Regression
MOS	Mean Opinion Scores
PCA	Principal Component Analysis
PDF	Probability Density Function
PESQ	Perceptual Evaluation of Speech Quality
PS	Power Spectrum
RGB	Red Green Blue
ROI	Region of Interest
SAAM	Semi Adaptive Appearance Models
SegSNR	Segmental SNR
SNR	Signal to Noise Ratio
STFT	Short-Time Fourier Transform
SVM	Support Vector Machine
TFGSC	Transfer Function Generalised Sidelobe Canceller
VAD	Voice Activity Detector
WSS	Weighted-Slope Spectral Distance
XM2VTSDB	The Extended Multi Modal Verification for Teleservices and Security Applications Database

Chapter 1
Introduction

Abstract Previous research developments in the field of speech enhancement (such as multi microphone arrays and speech enhancement algorithms) have been implemented into commercial hearing aids for the benefit of the deaf community. In recent years, electronic hardware has advanced to such a level that very sophisticated audio only hearing aids have been developed. It is expected that in the future, conventional hearing aids will be transformed to also make use of visual information with the aid of camera input, combining audio and visual information to improve the quality and intelligibility of speech in real-world noisy environments.

Keywords Audiovisual · Multimodal · Cognitive inspiration · Speech enhancement · Overview

1.1 Multimodal Speech Enhancement

Previous research developments in the field of speech enhancement (such as multi microphone arrays and speech enhancement algorithms) have been implemented into commercial hearing aids, and it is expected that in the future, conventional hearing aids will be transformed to also make use of visual information with the aid of camera input, combining audio and visual information to improve the quality and intelligibility of speech in real-world noisy environments. Amongst others, [1] investigated correlation between audio and visual features using Multiple Linear Regression (MLR), and expanded upon this to develop a visually derived Wiener filter for speech enhancement. The ultimate long term goal of the research presented in this book is to improve the lives of those who suffer from deafness. Even state-of-the-art modern hearing aids can fail to cope with rapid changes in environmental conditions such as transient noise or reverberation, and there is much scope for improvement. The work presented here aims to develop an initial cognitively inspired, autonomous, adaptive, and context aware multimodal speech enhancement framework, using fuzzy logic as part of the speech filtering process.

A. Abel and A. Hussain, *Cognitively Inspired Audiovisual Speech Filtering*,
SpringerBriefs in Cognitive Computation, DOI 10.1007/978-3-319-13509-0_1

1.2 Cognitively Inspired Intelligent Flexibility

The multimodal nature of both perception and production of human speech is well established. Speech is produced by the vibration of the vocal cords and the configuration of the vocal tract, which is composed of articulatory organs. Due to the visibility of some of these articulators (such as lips, teeth and the tongue), there is an inherent relationship between the acoustic and visible properties of speech production. The relationship between audio and visual aspects of speech perception has been established since pioneering work by Sumby and Pollack in 1954 [2], which demonstrated that lip reading improves the intelligibility of speech in noise when audiovisual perception is compared with equivalent audio-only perception. This was also confirmed by others [3], including in work by Summerfield in 1979 [4]. This work reported gains in the region of 10–15dB when compared to audio-only perception results [5], further demonstrated by the McGurk Effect [6], which provides a physical demonstration of the audiovisual relationship in terms of speech perception. This cognitive link between audio and visual information is further demonstrated in work concerning audio and visual matching in infants by Patterson and Werker [7, 8]. This correlation between audio and visual modalities can also be seen in studies of primates [9].

Further confirming the cognitive links between modalities, it has been shown that speech is perceived to sound louder when the listener looks directly at the speaker [10], as if audio cues are visually enhanced [11]. In addition, work by researchers including Kim and Davis [12], and Bernstein et al. [13], has shown that visual information can improve the detection of speech in noise [14]. Furthermore, work by Schwartz et al. [10] also investigated if visual cues present in speech information improve intelligibility, by using French vowels with very similar lip information and then dubbing different (but very similar) audio information over it. Despite the information not matching, a gain in intelligibility was identified when audiovisual information was used, suggesting very early audio and visual integration.

Studies have also shown that when informational masking (e.g. a competing speaker) is considered, visual information can have a dramatic effect, including research by Helfer and Freyman [15], and Wightman et al. [16]. An additional detailed discussion of audiovisual speech perception is presented in [17], and a further summary can be found in work by the authors in [18]. In addition, the correlation between audio and visual aspects of speech has been deeply investigated in the literature [19–21], and in work by the authors [22, 23], showing that facial measures provide enough information to reasonably estimate related speech acoustics.

The connection between modalities demonstrates the cognitive nature of hearing. The improvement found in perception and detection of speech with visual information, along with the McGurk Effect, shows that the process of hearing involves cognitive influence. In addition, the switch in attention focus to use varying amounts of visual information depending on relevance, and also the use of visual cues shows that there is a significant degree of processing in the brain.

There have been many different speech enhancement systems developed, both audio only [24, 25], and in recent years, multimodal [26, 27]. Often, these filtering systems are designed to perform best in very specific scenarios, but in a more realistic scenario, a degree of flexibility when it comes to speech processing is desirable. The potential flexibility of a multimodal speech enhancement system is of interest because in a real world environment, a number of conditions may vary. Different speech enhancement algorithms are suited to different conditions, and no single algorithm is ideal for use in all noise environments. Because of this, it is essential that a state-of-the-art multimodal speech filtering system is intelligent and sophisticated enough to take account of both acoustic and visual criteria to optimise speech filtering output.

References

1. I. Almajai, B. Milner, J. Darch, S. Vaseghi, Visually-derived Wiener filters for speech enhancement in *IEEE International Conference on Acoustics, Speech and Signal Processing, ICASSP 2007*, vol. **4**, pp. 585–588 (2007)
2. W. Sumby, I. Pollack, Visual contribution to speech intelligibility in noise. J. Acoust. Soc. Am. **26**(2), 212–215 (1954)
3. N.P. Erber, Auditory-visual perception of speech. J. Speech Hear. Disord. **40**(4), 481 (1975)
4. Q. Summerfield, Use of visual information for phonetic perception. Phonetica **36**(4–5), 314–331 (1979)
5. F. Berthommier, A phonetically neutral model of the low-level audio-visual interaction. Speech Commun. **44**(1), 31–41 (2004)
6. H. McGurk, J. MacDonald, Hearing lips and seeing voices. Nature **264**, 746–748 (1976)
7. M.L. Patterson, J.F. Werker, Two-month-old infants match phonetic information in lips and voice. Dev. Sci. **6**(2), 191–196 (2003)
8. M.L. Patterson, J.F. Werker, Matching phonetic information in lips and voice is robust in 4.5-month-old infants. Infant Behav. Dev. **22**(2), 237–247 (1999)
9. A.A. Ghazanfar, K. Nielsen, N.K. Logothetis, Eye movements of monkey observers viewing vocalizing conspecifics. Cognition **101**(3), 515–529 (2006)
10. J.-L. Schwartz, F. Berthommier, C. Savariaux, Seeing to hear better: evidence for early audio-visual interactions in speech identification. Cognition **93**(2), B69–B78 (2004)
11. K.W. Grant, P.-F. Seitz, The use of visible speech cues for improving auditory detection of spoken sentences. J. Acoust. Soc. Am. **108**, 1197 (2000)
12. J. Kim, C. Davis, Testing the cuing hypothesis for the AV speech detection advantage, in *AVSP 2003-International Conference on Audio-Visual Speech Processing* (2003)
13. L.E. Bernstein, S. Takayanagi, E.T. Auer Jr., Enhanced auditory detection with AV speech: perceptual evidence for speech and non-speech mechanisms, in *AVSP 2003-International Conference on Audio-Visual Speech Processing* (2003)
14. K.W. Grant, The effect of speechreading on masked detection thresholds for filtered speech. J. Acoust. Soc. Am. **109**, 2272 (2001)
15. K.S. Helfer, R.L. Freyman, The role of visual speech cues in reducing energetic and informational masking. J. Acoust. Soc. Am. **117**, 842 (2005)
16. F. Wightman, D. Kistler, D. Brungart, Informational masking of speech in children: auditory-visual integration. J. Acoust. Soc. Am. **119**, 3940 (2006)
17. D. Sodoyer, B. Rivet, L. Girin, C. Savariaux, J.-L. Schwartz, C. Jutten, A study of lip movements during spontaneous dialog and its application to voice activity detection. J. Acoust. Soc. Am. **125**(2), 1184–1196 (2009)

18. A. Abel, A. Hussain, Novel two-stage audiovisual speech filtering in noisy environments. Cogn. Comput. **6**, 1–18 (2013)
19. H. Yehia, P. Rubin, E. Vatikiotis-Bateson, Quantitative association of vocal-tract and facial behavior. Speech Commun. **26**(1–2), 23–43 (1998)
20. J. Barker, F. Berthommier, Estimation of speech acoustics from visual speech features: a comparison of linear and non-linear models, in *AVSP'99-International Conference on Auditory-Visual Speech Processing* (1999)
21. I. Almajai, B. Milner, Maximising audio-visual speech correlation, in *Proceedings AVSP* (2007)
22. S. Cifani, A. Abel, A. Hussain, S. Squartini, F. Piazza, An investigation into audiovisual speech correlation in reverberant noisy environments, in *Cross-Modal Analysis of Speech, Gestures, Gaze and Facial Expressions: COST Action 2102 International Conference Prague, Czech Republic, 15–18 October 2008 Revised Selected and Invited Papers*, vol. 5641 (Springer, 2009), pp. 331–343
23. A. Abel, A. Hussain, Q. Nguyen, F. Ringeval, M. Chetouani, M. Milgram, Maximising audio-visual correlation with automatic lip tracking and vowel based segmentation, in *Biometric ID Management and Multimodal Communication: Joint COST 2101 and 2102 International Conference, BioID_MultiComm 2009, Madrid, Spain, 16–18 September 2009, Proceedings*, vol. 5707, (Springer, 2009), pp. 65–72
24. T. Van den Bogaert, S. Doclo, J. Wouters, M. Moonen, Speech enhancement with multichannel Wiener filter techniques in multimicrophone binaural hearing aids. J. Acoust. Soc. Am. **125**, 360–371 (2009)
25. J. Li, S. Sakamoto, S. Hongo, M. Akagi, Y. Suzuki et al., A two-stage binaural speech enhancement approach for hearing aids with preserving binaural benefits in noisy environments. J. Acoust. Soc. Am. **123**(5), 3012–3012 (2008)
26. R. Goecke, G. Potamianos, C. Neti, Noisy audio feature enhancement using audio-visual speech data, in *IEEE International Conference on Acoustics, Speech, and Signal Processing, 2002. (ICASSP'02)*, vol. 2, (2002), pp. 2025–2028
27. I. Almajai, B. Milner, Effective visually-derived Wiener filtering for audio-visual speech processing, in *Proceedings of the Interspeech*, Brighton, UK (2009)

Chapter 2
Audio and Visual Speech Relationship

Abstract This chapter presents a summary of the general research domain, and also the relationship between audio and visual aspects of speech. The background to human speech production is briefly discussed, along with a definition of several speech phenomena relevant to this topic, namely the Cocktail Party Problem, McGurk Effect, and Lombard Effect, and also briefly discusses audiovisual speech correlation. The relationship between audio and visual speech features has been deeply investigated in the literature, and a summary of this has been provided, with a particular focus on recent work by [3], however, very little work (to the knowledge of the author) has been carried out into audiovisual speech correlation with noisy speech. This chapter presents a brief background of human speech production, followed by a summary of several speech phenomena relevant to this research domain in Sect. 2.2. A review of audiovisual correlation research is then described in Sect. 2.3.

Keywords Audiovisual · Multimodal · Speech relationship · Correlation

2.1 Audio and Visual Speech Production

2.1.1 Speech Production

The multimodal nature of human speech is established. This use of multiple modalities in speech involves production as well as perception; indeed, speech is produced by the vibration of the vocal cords and the configuration of the vocal tract. Since some of these articulators are visible, there is an inherent relationship between the acoustic and visual aspects of speech. Moreover, the well-known McGurk Effect (summarised later in this chapter) empirically demonstrates that speech perception also makes use of multiple modalities. The high degree of correlation between audio and visual speech has been deeply investigated in the literature, with work by [1–3], showing that facial measures provide enough information to reasonably estimate related speech acoustics. This is a subject area which has been researched in depth, and there are many detailed summaries in the literature, including by [4], and [5], and this section contains a brief summary. In speech production, the shape of the

A. Abel and A. Hussain, *Cognitively Inspired Audiovisual Speech Filtering*, SpringerBriefs in Cognitive Computation, DOI 10.1007/978-3-319-13509-0_2

vocal tract is decided by the articulators, such as the lips, tongue and teeth. Depending on the vibration of the vocal cords, speech can be defined as either voiced or unvoiced. Voiced sounds include all vowels, and some consonants, caused by the vibration of the vocal cords. Unvoiced sounds do not involve vibration, and involve airflow passing through an opening in the vocal cords, with a noise being produced by constriction in the vocal tract. There are also plosive sounds, which are made by closing the lips, allowing air pressure to build, and then opening them again.

2.1.2 Phonemes and Visemes

Speech can be divided by artificial notation, for both audio (phonemes) and visual (visemes). A full detailed description can be found in work by [4]. Phonemes are very basic audio speech units [6]. These are broken down from individual words, and there are many different notations used to represent phonemes, depending on the language and source. With regard to British English, phonemes can be broken down into four main categories, vowels, consonants, diphthongs, and semivowels. To describe individual visual speech units, visemes are used. However, these do not map exactly to phonemes due to the nature of the articulators used in speech production (i.e. the lips are always visible, whereas others like the tongue are only visible intermittently). There are a variety of ways in which these have been mapped. This is considered to be outside of the scope of this work, and so has not been covered here.

2.2 Multimodal Speech Phenomena

2.2.1 Cocktail Party Problem

The Cocktail Party problem was first defined by [7] in 1953, and describes the ability of human listeners to be able to listen to a single speech source while unconsciously filtering out irrelevant background information such as music or competing speech sources. This phenomenon was named after the scenario of two people being able to maintain a conversation while ignoring the sound of a lively party with a myriad of competing speakers and other background noise. Despite being defined in 1953, there is no definitive explanation of this effect, and it represents a very active field of research. A detailed review of this effect and its technological application has been carried out by [8].

This effect is of relevance to this field of research because it represents a major challenge for those that suffer from hearing loss. The Cocktail Party effect seems to be binaural, with implications for sound localisation, and those with hearing aids or with 'unbalanced hearing' can particularly struggle to cope with noisy environments.

There has also been much research into solving this problem using speech filtering technology. An example of this is BSS, with many recent examples of research aiming to solve this problem such as work by [9].

2.2.2 *McGurk Effect*

The McGurk Effect was first reported in 1976 by [10]. The significance of this effect with regard to the work discussed here is that it serves as a physical demonstration of the relationship between hearing and vision in terms of speech perception. Essentially, when a video is played of the syllable 'ga', the viewer recognises it correctly. However, when the video is dubbed with the audio for 'ba', the viewer hears a third syllable, 'da'. The conflict between the sound and the visual effect of the mouthing of a syllable results in the viewer hearing something different.

2.2.3 *Lombard Effect*

The Lombard Effect, discovered in 1909 [11], describes the tendency of speakers to attempt to improve the audibility of their voice in loud environments by involuntarily increasing their vocal effort. Research has shown that in very noisy environments, speakers can change their pitch, frequency, duration, and their method of vocalisation in order to be heard more clearly. This is not a voluntary phenomenon, and an example of this effect is found in deaf people, who often speak loudly due to their hearing loss, in order to hear their own voice more clearly. Examples can be found in work by [12], who presented the AVICAR corpus. This effect has implications for speech processing research, for example, many visual speech estimation or recognition models are trained only on clean speech, and so performance may therefore be poor in noisy environments.

2.3 Audiovisual Speech Correlation Background

There is an established relationship between audio and visual aspects of both speech perception and production. This can be expressed as the correlation between audio and visual speech features. There are various methods of calculating correlation such as MLR [3, 13] and CCA [14, 15], and a significant quantity of research has been published investigating audiovisual speech correlation. Relatively early examples include [1], who used LSP audio features and examined the correlation between these and 3D marker points (calculated using infrared LEDs physically placed on the speaker), and [2], who investigated correlation between face movement (not confined to lip features alone) and LSP audio features, and found a slightly lower correlation.

More recent work makes use of CCA [14] to investigate audiovisual correlation. CCA was used by [15] to analyse the linear relationships between multidimensional audio and visual speech variables by attempting to identify a basis vector for each variable, that then produces a diagonal correlation matrix. CCA maximises the diagonal elements of the correlation matrix. Sargın et al. [15] made use of the commonly used MFCC technique for audio features, and then compared the correlation when using three different types of visual features, 2D-DCT features taken from image frames, 2D-DCT features taken from optical flow features, and predefined lip contour co-ordinates.This work concluded that individual correlation was greatest with the 2D-DCT of optical flow vectors. This was then extended by [16] in later work. The correlation work performed by [15] served as a background to work by the authors in [17].

Recent work by [3] also investigated the degree of audiovisual correlation between multiple audio and visual features, when considering different vector dimensionalities of each component. Almajai and Milner [3] used filterbank vectors and the first four formant frequencies (the most significant distinguishing frequency components of human speech) as audio features, and three different visual features, 2D-DCT, Cross-DCT, and AcAMs. AcAMs are a commonly used approach for feature extraction, first developed by [18], and operate by building statistical models of shape and appearance, based on a training set. This was expanded upon by the author in [13] and used as a basis for new experiments into audiovisual speech correlation using MLR.

The following conclusions were drawn by [3]. As the size of the visual vector increased, correlation increased. However, this initially increased rapidly until a dimensionality of 24, and then stabilised. Increasing the size of the vector further was not found to make a particularly significant difference. On the other hand, decreasing the size of the filterbank vector increased correlation. This was argued to come at the cost of potentially useful speech information. It was also concluded that Cross-DCT resulted in a lower correlation than using 2D-DCT or AcAMs. Due to the simplicity of using 2D-DCT visual features rather than AcAMs, it was recommended that the 2D-DCT visual feature technique be used for further research. Additionally, [3] found after an investigation of phoneme specific versus global speech correlation that phoneme specific correlation analysis also resulted in an increased correlation.

2.4 Multimodal Correlation Analysis

In addition to the audiovisual correlation work presented in the literature, additional correlation research has been published by the authors [13, 17] investigating the relationship between audio and visual elements of speech. The research presented in [13] is briefly summarised in this section. This work also serves as a comparison and validation of similar preliminary work by [3], as it uses a different corpus but with some overlap in the techniques used.

2.4.1 Correlation Measurement

To carry out the audiovisual correlation reported in [3] and by the authors in [13], MLR is used. This is a multivariate approach that assesses the relationship between audio and visual vectors [19]. In the analysis summarised in this chapter, experiments have been carried out using an audio frame of 25 ms and a video frame of 100 ms, based on the parameters also used by existing work in the literature [3]. This implies that the same visual features are used for four consecutive audio frames. For a speech sentence, each component $F_a(l, j)$ of the audio feature vector is predicted by means of MLR, using the entire visual feature vector $F_v(l, q)$, where l is the time-frame index. This means that using $Q + 1$ regression coefficients $\{b_{j,0}, ..., b_{j,q}, ..., b_{j,Q}\}$ the jth component of the audio feature vector can be represented by the visual feature vector $F_v(q) = [F_v(l, 0), ..., F_v(q), ..., F_v(Q - 1)]$,

$$\hat{F}_a(l, j) = b_{j,0} + b_{j,1} F_v(l, 0) + \cdots + b_{j,Q} F_v(l, Q - 1) + \epsilon_l \tag{2.1}$$

With ϵ_l representing an error term. The multiple correlation between the jth component of the audio feature vector and the visual vector, calculated over L frames, is given by R_s, and is found by calculating the squared value:

$$R_s(j)^2 = 1 - \frac{\sum_{l=0}^{L} \left(F_a(j) - \hat{F}_a(j)\right)^2}{\sum_{l=0}^{L} \left(F_a(j) - \bar{F}_a(j)\right)^2} \tag{2.2}$$

$\bar{F}_a(j)$ represents the mean of the jth component of the audio feature vector. In this work, the single correlation value R is found by calculating the mean of each jth component of R_s. By this, we mean that R_s returns a vector of correlations, with each value representing the correlation of one audio component to the entire visual vector, and the mean of R_s produces the single correlation value R, which represents the correlation of the entire audio vector to the entire visual vector.

2.4.2 Multimodal Correlation Analysis Results

2.4.2.1 Comparison of Visual Feature Extraction Techniques

One conclusion of the work reported in [13] was a comparison of visual feature extraction techniques, most notably, comparing Cross-DCT and 2D-DCT. For a matrix of pixel intensities, to calculate the 2D-DCT a 1D-DCT is applied to each row of the matrix, and then to each column of the result. Thus, the transform is given by the DCT matrix. This is described in more depth in Chap. 4. Cross-DCT consists of taking only the central horizontal row and vertical column of the matrix of pixel intensities and then applying 1D-DCT to each vector. This theoretically contains adequate information for lip reading because the former captures the width of the mouth while the latter captures the height of the mouth. The two vectors are truncated and concatenated to get the visual feature vector. However, work by the authors in [13] found that in a comparison of sentences from the VidTIMIT corpus, the audiovisual correlation found when comparing MFCC correlation to 2D-DCT visual features was greater in all tested cases than for the equivalent correlation when using Cross-DCT, therefore justifying the use of 2D-DCT.

2.4.2.2 Maximising Audiovisual Correlation

Experiments to investigate the ideal audio and visual feature vector dimensionalities to use for performing multimodal correlation analysis were reported by the authors in [13]. This was done by adding white noise to 16 sentences from the VidTIMIT corpus in order to produce noisy speech with a SNR of -3 dB, and making use of a beamformer (which will be discussed in depth in Chap. 4) to remove the added noise and produce enhanced speech. In the experiments summarised here, four microphones were used in a simulated room environment. This configuration was subsequently used in additional research, and is described in Chap. 4 . The correlation was found by performing MLR when varying the MFCC and 2D-DCT vector sizes. As reported in [13], increasing the size of the visual vector increases the correlation of the enhanced results, and reducing the size of the audio vector produces a similar effect, peaking at very high (70) visual and very low (3) audio vector dimensionalities, which initially seems to differ from results found in the literature [3]. Almajai and Milner [3] found that the increase in correlation levelled off beyond a visual vector size of 30. However, a visual comparison of this to the equivalent results for correlation with noisy (unfiltered) speech shows that they have a similar shape, and where the enhanced correlation is very large; the noisy correlation is also very large. Therefore, it becomes important to find the audio and visual vector sizes that maximise the difference between noisy and enhanced speech.

When taking the difference between noisy and enhanced speech correlation plotted against varying audio and visual vector sizes for a mean of 16 sentences respectively, the authors found in [13], that with a very small visual vector, the difference in audiovisual correlation is very small, and that an initial increase results in an increased difference between noisy and enhanced correlation. However, this increase tails off when the visual vector is increased above thirty, showing that there is no significant gain to be achieved from increasing this above thirty, matching and validating results found in the literature by [3]. However, the individual correlation results were found to be lower to those reported by others. It is hypothesised that this is because the work described in this chapter makes use of a different corpus (VidTIMIT), which contains a great deal of background noise, which has the effect of producing lower levels of audiovisual correlation than would be found when using a cleaner multimodal speech corpus.

Additionally, [13] also found that increasing the size of the MFCC dimensionality results in a lower difference in audiovisual correlation and that the highest difference is found with an audio vector size of less than five. However, this is not a practical value to use. A very low MFCC dimensionality contains less spectral information about the input speech, and a compromise between maximising correlation and feasibility for complex speech processing use has to be found. These experiments are fully described in [13].

References

1. H. Yehia, P. Rubin, E. Vatikiotis-Bateson, Quantitative association of vocal-tract and facial behavior. Speech Commun. **26**(1–2), 23–43 (1998)
2. J. Barker, F. Berthommier, Estimation of speech acoustics from visual speech features: a comparison of linear and non-linear models, in *AVSP'99-International Conference on Auditory-Visual Speech Processing*, (1999)
3. I. Almajai, B. Milner, Maximising audio-visual speech correlation, in *Proceeding AVSP*, (2007)
4. I. Almajai, Audiovisual speech enhancement. Ph.D. thesis, University of East Anglia, (2009)
5. F. Owens, P. Lynn, *Signal processing of speech* Macmillan New Electronics, (1993)
6. L. Rabiner, R. Schafer, *Digital Processing of Speech Signals*, vol. 100 (Prentice–hall Englewood Cliffs, New Jersey, 1978)
7. E. Cherry, Some experiments on the recognition of speech, with one and with two ears. J. Acoust. Soc. Am. **25**(5), 975–979 (1953)
8. S. Haykin, Z. Chen, The cocktail party problem. Neural Comput. **17**(9), 1875–1902 (2005)
9. B. Rivet, Blind non-stationnary sources separation by sparsity in a linear instantaneous mixture, in *Independent Component Analysis and Signal Separation*. Lecture Notes in Computer Science, vol. 5441, ed. by T. Adali, C. Jutten, J. Romano, A. Barros (Springer, Berlin, 2009), pp. 314–321
10. H. McGurk, J. MacDonald, Hearing lips and seeing voices. Nature **264**, 746–748 (1976)
11. E. Lombard, Le signe de l'elevation de la voix. Ann. Maladies Oreille Larynx Nez Pharynx **37**(101–119), 25 (1911)
12. B. Lee, M. Hasegawa-johnson, C. Goudeseune, S. Kamdar, S. Borys, M. Liu, T. Huang, AVICAR: Audio-Visual Speech Corpus in a Car Environment, in *Proceeding of the International Conference Spoken Language* (Jeju, Korea, 2004), pp. 2489–2492, Citeseer
13. S. Cifani, A. Abel, A. Hussain, S. Squartini, F. Piazza, An investigation into audiovisual speech correlation in reverberant noisy environments, in *Cross-Modal Analysis of Speech, Gestures, Gaze and Facial Expressions: COST Action 2102 International Conference Prague, Czech Republic, October 15-18, 2008 Revised Selected and Invited Papers*, vol. 5641, (Springer-Verlag, 2009) pp. 331–343,
14. H. Hotelling, Relations between two sets of variates. Biometrika **28**(3/4), 321–377 (1936)
15. M. Sargın, E. Erzin, Y. Yemez, A. Tekalp, Lip feature extraction based on audio-visual correlation, in *Proceeding EUSIPCO*, vol. 2005, (2005)
16. M. Sargin, Y. Yemez, E. Erzin, A. Tekalp, Audiovisual synchronization and fusion using canonical correlation analysis. IEEE Trans. Multimed. **9**(7), 1396–1403 (2007)
17. A. Abel, A. Hussain, Q. Nguyen, F. Ringeval, M. Chetouani, M. Milgram, Maximising audiovisual correlation with automatic lip tracking and vowel based segmentation, in *Biometric ID Management and Multimodal Communication: Joint COST 2101 and 2102 International Conference, BioID_MultiComm 2009, Madrid, Spain, September 16-18, 2009, Proceedings*, vol. 5707, (Springer 2009) pp. 65–72,

18. T. Cootes, G. Edwards, C. Taylor, Active appearance models, in *Computer Vision-ECCV 98 (1998)*, pp. 484–498
19. S. Chatterjee, A. Hadi, *Regression Analysis by Example*, 4th edn. (Wiley, 2006)

Chapter 3
The Research Context

Abstract This chapter presents a literature review that places the research proposed in this book in context, building on the background presented in the previous chapters. Firstly, the overall speech processing domain is briefly discussed. This review presents examples of listening devices using directional microphones, array microphones, noise reduction algorithms, and rule based automatic decision making, demonstrating that the multimodal two stage framework presented later in this book has established precedent in the context of real world hearing aid devices. The other significant aspect vital to the research context of this work is the field of audiovisual speech filtering. This chapter presents a review of multimodal speech enhancement, with a discussion of the initial early stage audiovisual speech filtering systems in the literature, and the subsequent development and diversification of this field. A number of different state of the art speech filtering systems are examined and reviewed in depth, particularly multimodal beamforming and Wiener filtering. Finally, several audiovisual speech databases are evaluated.

Keywords ROI · Visual tracking · Speech enhancement · Multimodal · Corpus · Review

3.1 Application of Speech Processing Techniques to Hearing Aids

This work considers speech filtering from the point of view of potential application to hearing aids for the benefit of users with deafness. However, this is very much a long term focus, with the work presented in this book focusing exclusively on early stage software development. However, it is considered appropriate to provide an overview of some features of modern hearing aids. Much of the content in this section is adapted from a detailed review by [1]. The review by [1] states that hearing aid technology can be divided into two categories, directional microphones and noise reduction algorithms.

A. Abel and A. Hussain, *Cognitively Inspired Audiovisual Speech Filtering*,
SpringerBriefs in Cognitive Computation, DOI 10.1007/978-3-319-13509-0_3

3.1.1 Directional Microphones

With regard to directional microphones, the most relevant topics include first order directional microphones, adaptive directional microphones, second order directional microphones, and assistive array microphone devices. Overall, directional microphones are an established technology that has been around since the 1970s [1]. They operate on the premise that speakers are more likely to be located to the front of the listener, and so directional microphones are designed to be more sensitive to sounds arriving from the front of the speaker.

First Order Directional Microphones
The most common type of directional microphone processing was identified by [1] as being first order directional microphones. These are designed to direct the focus of microphone sensitivity to sounds coming from the front of the listener and reduce sensitivity to sounds arriving from the side or rear of the listener. Hearing aids equipped with this technology can be designed to use single or dual microphones. In the single microphone configuration, a hearing aid has a microphone with two ports, anterior and posterior. Sound entering the posterior port is delayed and subtracted from the input to the anterior port. This delay is determined by physical factors such as the distance between the two microphone ports. These have been superseded in recent years by dual microphone hearing aids, which work in a similar fashion. These have two omnidirectional microphones, with an anterior and posterior microphone. The microphone inputs are combined with 'delay and subtract' processing, similar to that discussed for single microphone hearing aids. The difference is that the delay and therefore the directional focus is software based and can be adjusted and programmed with signal processing algorithms [2].

With regard to performance, it is consistently found in laboratory testing that the use of directional microphones can lead to improved results [3]. The more focused the directionality is, the better the results. Optimum performance was produced when there were less discrete noise sources, and less reverberation. This is because a large room produces a lot of reverberation and these echoes reduce the effectiveness of directional focus. However, in actual practical usage, many users do not perceive the benefit of directional microphones. There are several reasons for this. Firstly, the desired signal is not always located directly to the front of the listener, many times; the speaker will be at an angle to the listener [4]. There are also a wide variety of reverberation environments in real life, with many unsuited to directional microphones, and so the large improvements seen in laboratory experiments do not always translate well into practical use.

Another issue is that research found that the majority of hearing aid users simply do not use the directional setting and preferred to use an omnidirectional mode at all times [5, 6]. There are significant limitations to using directional microphones in many environments. For example, in quiet environments, when there is wind noise present, or when reverberation is at a significant level, omnidirectional microphones are suggested, and many users prefer to use this setting at all times. So although very commonly used, directional microphones have significant limitations.

Adaptive Directional Microphones

To improve upon the limitations of directional microphones, adaptive directional microphone techniques have been developed. While first order directional microphones assume a fixed source location at the front of the listener, adaptive microphones do not make this assumption, and are designed with the aim of maximising sensitivity in the direction of the dominant source location and minimising sounds originating from the opposite direction to this source. This makes adaptive dual microphone hearing aids theoretically more suited to real world environments.

In terms of functionality, adaptive microphone hearing aids function in a similar way to the first order directional microphones described above, but the direction and microphone settings are not fixed. These settings (such as the posterior delay and omni or uni directional mode) can be automatically changed with sophisticated signal processing. Different manufacturers use a range of proprietary methods, which are not generally publicly disclosed, but [1] provides a general overview of the functionality of adaptive microphones. In terms of results, [1] concludes that adaptive directional microphones have not been found to deliver a worse performance than fixed first order directional microphones in many scenarios, but in scenarios with a very narrow angle of speech, can produce improved results. More detailed results are given in [7].

Second Order Directional Microphones

An alternative to first order directional microphones is to use second order directional microphone hearing aids. First order microphones are normally found to result in a 3–5 dB SNR improvement. However, for those with a significant hearing loss, this improvement may not be noticeable. Second order directional microphones utilise more than two microphones, but otherwise use similar delay and subtract processing techniques. The downside to this approach is that there is a lot of low frequency roll off, which is difficult to amplify without having an impact on the amplification of internal noise. Bentler et al. [8] tested one such system, the Siemens Triano, and found only small benefits. This design of hearing aid is reported by [1] as being relatively uncommon.

Assistive Listening Array Microphones

The previous sections discussed hearing aids that delivered several decibels of SNR improvement. For those with more than 15 dB of hearing loss though, the gains are still not sufficient to adequately increase the SNR of a desired speech source in difficult environments. The traditional solution for this is to use a FM radio microphone system. This consists of a microphone placed close to the mouth of a speaker (such as a wearable clip on microphone), which is then transmitted via radio to a receiver worn by a listener. This delivers very good performance, but comes at the cost of removing almost all background noise. It is also limited to a single speaker. One alternative to this traditional approach is to use a microphone array.

A multiple microphone array consists of a series of linked microphones, with the inputs combined to provide directional focus. Delay and sum processing is used with each microphone to increase the directional effect to a greater extent than for conventional hearing aids. The array input then overrides the user's hearing aid input. These arrays can be hand held (i.e. a device that the listener points at the desired target), or head worn [9]. In terms of results, [10] reported a 7–10 dB improvement, which was backed up by [11]. However, although a 7–10 dB gain is reported, a traditional FM transmitter system still has a much greater level of improvement, and is still the recommended solution for noisy environments with a single speaker.

3.1.2 Noise Cancelling Algorithms

While directional microphones take advantage of the spatial differences between speech and noise sources, noise reduction algorithms aim to exploit the spectral and temporal differences between speech and noise. Noise reduction relies on speech filtering software algorithms with decision rules used to decide the appropriate level of filtering. These rules rely on the input from various detectors such as wind noise, signal level, and modulation detectors. Again, the exact configuration and use of detectors tends to be proprietary, although the use of a modulation detector is considered to be standard practice to detect the presence of speech [12]. Chung [1] distinguishes between algorithms that detect the specific modulation of speech (multichannel adaptive noise reduction algorithms), and those that detect co-modulation (synchrony detection noise reduction algorithms).

Multichannel Adaptive Noise Reduction Algorithms

Multichannel adaptive noise reduction algorithms aim to reduce noise interference at frequency channels where noise is dominant. These algorithms are most effective when there is a significant spectral difference between speech and noise, but suffer if the noise source is speech from a competing speaker. In general, these algorithms are described by [1] as having three stages. First, signal detection and analysis is carried out. This is then followed by the application of decision rules, and finally, appropriate gain reduction (adjustment of the ratio of output to input) is carried out.

The first stage, signal detection and analysis, is similar to that for directional microphones, in that a variety of detectors such as wind and modulation detectors are used to analyse the input signal. After the initial input, the second phase is the application of decision rules. Again, this varies between individual hearing aids, and [1] reports that a range of factors such as the level of input signal, the SNR at individual channels, and the type of noise reduction programmed for the individual user during the hearing aid fitting process are considered. The outcome of the decision rules is to apply gain reduction at the appropriate frequency channels. This is not an exact science, and different balances between listening comfort and performance are found on different hearing aids [13].

Overall results of these algorithms are reported as being mixed. Although positive results have been reported in the literature by [14, 15], others have reported no significant improvements [16]. Generally, the bigger the difference between speech and noise, the greater the benefits of this form of noise reduction.

Synchrony Detection Noise Reduction Algorithms
Synchrony detection noise reduction algorithms take advantage of the detection of co-modulation in speech to distinguish between speech and noise [17]. These algorithms essentially detect the fast modulation of speech across frequency channels. The signal detector monitors high frequency parts of the incoming signal and looks for the high frequency spectral spikes that indicate co-modulation. If these rapid bursts of speech energy are detected then it is assumed that speech is present and gain levels are kept at the default level. However, if co-modulation is not detected, then it is assumed that speech is not present and the overall gain of the hearing aid is gradually reduced [18].

Chung [1] reports that this algorithm is less useful when the noise source is competing speech because this speech also has co-modulation, and that this algorithm can be combined as part of an overall noise reduction package combining both multichannel and synchrony detection algorithms, rather than being used individually.

3.2 Audiovisual Speech Enhancement Techniques

3.2.1 Background

Given the audio and visual speech relationship described previously, and recent correlation research, it was obvious that the concept of audio-only speech enhancement systems would be extended to become multimodal. A pioneering multimodal speech enhancement technique was proposed by [19]. This was, to the knowledge of the author, the first example of a functioning multimodal speech enhancement system. This approach made use of the height, width, and area of the lips, recorded using blue lipstick and chroma key technology to record this information and exclude other details, and made use of simple linear regression to estimate a Wiener filter to enhance speech contaminated with white noise. Results performed on selected data from vowel/consonant/vowel sequences showed an improvement over similar audio-only approaches, but were limited due to the linear nature of the filter. A non-linear artificial neural network utilising MLP was proposed, and this more complex filter was found to produce improved results. This early work demonstrated the potential of multimodal enhancement, and was developed further by [20], who proposed a linear mean square error estimation method. Deligne [21] demonstrated a non-linear approach for speech enhancement, AVCDCN, which was an extension of an audio-only CDCN [22, 23] approach. The authors found that limited experiments using the same noise for testing as was used in training showed improved results when the audiovisual approach was used rather than audio-only. Since this early work, there have been many recent advances in this research field. Three of the most relevant state of the art developments that particularly build on this early work are discussed in depth here.

3.2.2 Audiovisual Blind Source Separation

3.2.2.1 Summary of Work and Previous Papers

A range of authors including [24] (also [25–28]) have been involved in the development of multimodal BSS systems, which aim to filter a speech source from a noisy convolved speech mixture. This expands upon previous related work by these authors, and is an extension of audio-only BSS solutions. BSS, first proposed by [29] (also [30]) was designed with the aim of separating individual speech sources from a mixture of competing speakers. The problem of separating speech mixtures and recovering individual speech sources is one that is of great interest to researchers due to its relevance to the Cocktail Party Problem [31]. It is a difficult problem to tackle as real world speech mixtures are convolved. By that, it is meant that the sources are mixed, reflected off of different surfaces in the speech environment, and weakened before they are picked up by the microphones. The 'blind' aspect of the name refers to a lack of knowledge regarding information about the number of sources and the mixing matrix.

The research presented in this section attempts to use visual information to tackle these limitations. A detailed review is presented in a recent paper by [24], and is summarised here. In all of the examples presented, visual information is used to assist with the estimation of the permutation and diagonal matrices. This work builds upon previous related work by [29, 32–34], which has experimented with the fusion of audio and visual information. Initial multimodal source separation work [33, 34], focused on maximising an audiovisual statistical model in order to extract the correct signal. This was found to be computationally expensive, especially when convolved speech mixtures were considered. One approach considered was to maximise the relationship between audio source information and lip movement with a statistical model, as detailed further in work by [25, 26]. Another approach to solve the permutation indeterminacy problem is arguably more computationally efficient and makes use of a VAD to identify silent periods in speaker utterances [27, 28], and so extract the correct source at all frequency bins.

3.2.2.2 Key Output

Research by [35] builds on the concept of using visual information to solve permutation indeterminacies, and develops an audiovisual beamforming approach to solve the problem of source separation in an environment consisting of a mixture of overlapping moving speech sources. The work in [35] utilises a simpler approach that uses speaker tracking to identify source locations, and then uses these coordinates for beamforming.

One significant aspect of this system is the 3D visual tracking approach utilised. Speaker tracking is used to identify visual data and makes use of state of the art techniques. Firstly, some assumptions are made. It is assumed by [35] for the purposes of their paper that a full face image of each speaker is visible at all times, and that a geometric cue (i.e. the centre of the face) is available. The experiments are performed in a simulated office environment (a small room), using two high quality cameras, mounted above head height and synchronised using an external hardware trigger module. This provides a high vantage point, and the input from the two cameras is used to convert a two dimensional view of the room to 3D. To carry out the tracking, a Viola-Jones face detector is used [36]. The tracking stage ultimately calculates the 3D position and velocity of each speaker, which is used in the source separation stage.

After receiving visual information, the second stage is to perform source separation. In this work, the authors assume that noise is either non-existent, or considered to be a separate source, and that the number of input sources is equal to the number of desired outputs. The visual information is first used to determine whether the sources are moving or stationary. If the sources are considered to have been stationary for at least two seconds, then IIFastICA is used to separate the sources. This uses the estimated FIR filter and whitening [37] to initialise the FastICA algorithm [38]. If however, the sources are determined to be moving, beamforming is used, with the aid of visual information. See [35] for more information.

In the experiments reported by [35], the system is evaluated with a room containing two speakers, with audio recorded at 8 kHz and video at 25 Hz, and audio and visual data manually synchronised. Initially, the 3D tracking was evaluated, and was found to be accurate and effective. The angle-of-arrival data (using visual positioning) was also found to be correct with regard to the experimental data. With regard to the source separation problem, various system configurations (both audio-only and audiovisual) were tested by [35], and they found that the use of visual information improved overall results and algorithm performance.

3.2.2.3 Strengths

There are a number of strengths of this work. Firstly, this research demonstrates the value of using visual information as part of a speech filtering system. It builds on prior work and shows an improvement on existing audio-only techniques (for example, using audio-only geometric information for beamforming initialisation). The results demonstrate that the proposed multimodal system is effective. The recent state of the art work discussed above makes use of modern 3D tracking technology and displays a nuanced approach to speech processing. The type of source separation performed varies depending on environmental conditions (the movement of the speech sources), showing an intelligent use of multimodal information.

3.2.2.4 Limitations

There are a number of practical limitations to this work. Firstly, there are a number of assumptions made. It is assumed that the room being used is small enough to keep the reverberation level low, and it is also assumed that good visual information is visible, with a good quality full facial image available for each speaker at all times. This is adequate for the limited experiments discussed in this paper, but in a practical real world environment, these conditions are unlikely to be met. The cameras and microphones are also fixed in position, with the microphone array in the centre of the room at all times, and the cameras mounted above head height. This is adequate for simulations, but may produce poor results if experiments are extended to a more realistic environment with regard to hearing aid wearers. A hearing aid user would not be expected to remain stationary, and this means that the cameras and microphone would be mobile, making calculations such as the angle of arrival of different speech sources and accurate 3D tracking much more difficult. The system is also aimed specifically at solving the source separation problem, with noise that does not originate directly from a competing speech source not considered.

3.2.3 Multimodal Fragment Decoding

3.2.3.1 Summary of Work and Previous Papers

Multimodal speech fragment decoding, developed by [39], was designed to improve on existing audio-only fragment decoding techniques [40]. The primary problem that this system attempts to tackle is that of speech recognition in environments where the speech source is obscured by a competing simultaneous speech source, and this research is particularly focused on the problem of masking.

There are two types of masking that represent a problem for speech recognition systems. The first is energetic masking, which occurs when the speech energy of the masker is greater than the energy of the speech source. An example of this is when a vowel from a competing masking speaker obscures an unvoiced part of the target speech source. The second type of masking, informational masking, is more challenging. This is when it is unclear which part of the noisy speech input signal is dominated by the target speaker and which by a masking speaker. This research aims to tackle this by using visual information to identify the target speaker with greater accuracy.

Source separation research is inspired by ASA [41, 42], which is the process by which humans organise sounds into meaningful elements. The concept behind speech fragment decoding is that in a noisy speech mixture, there exist elements within this mixture (in the spectral-temporal domain) where speech energy is concentrated sufficiently to ensure that noise source energy has a negligible effect. There are two elements to the fragment decoding process. The first is the generation and identification of spectral temporal fragments, i.e. those fragments which are dominated by one single source, either target or masking source. These fragments are automatically labelled and are then used to create a segregation (represented by a binary mask) of these labelled fragments. A segregation hypothesis is then searched for, where the noisy input is then matched to statistical models of clean speech, with missing data speech recognition performed by matching fragments to HMMs (trained on clean speech), with the type of processing dependent on the labelling, in order to produce the best matching word sequence from the noisy input mixture. There are practical limitations to an audio-only approach though. While it is possible to determine whether a fragment is dominated by a single source, it can be difficult to determine whether that source is the target speaker or background noise.

3.2.3.2 Key Output

State-of-the-art multimodal work carried out using this technique [39, 43] utilises visual information alongside audio to assist with the identity of fragments dominated by the target speaker. In the case of labelling appropriate fragments as dominated by target or noise, visual information helps to increase the accuracy of this. In the simplest case, visual features can determine the likelihood of the target speaking or being silent at a given point in time. Trained audiovisual models can identify audio fragments that match the equivalent visual information well, increasing the accuracy of fragment labelling.

To extract the audio features, the input signal is passed through a 64-channel filterbank, and temporal difference features are computed with 5-frame linear regression and added to the 64-channel filterbank output to create an audio vector with a dimensionality of 128. The visual features take the form of 2D-DCT features extracted from the lip region of the target speaker, with 36 (6 by 6) low-order coefficients extracted. Temporal difference features are added to these to create a vector with a dimension of 72. As the video is recorded at 25 fps, this is up-sampled to 100 fps to match the audio vector.

The research presented in this work takes the audio and visual inputs and combines them in a variety of ways using HMMs, so both early stage feature fusion (concatenating audio and visual vectors before processing), and decision fusion (visual and audio computed separately and then merged) approaches are considered. In terms of spectral temporal fragment identification, the authors describe visual features as a form of 'scaffolding' that supports the fragment identification process. This research required the training of HMMs for each speaker tested.

The authors compared this approach to a similar audio-only fragment decoding approach using sentences from the GRID Corpus [44], and found that while the audio-only approach produced similar results as the audiovisual system with an SNR of +6 dB, as the SNR decreased to −9 dB the audio-only approach produced a much steeper drop off in performance, with the audiovisual approach performing significantly better. This shows the benefits of a multimodal approach.

3.2.3.3 Strengths

The most important thing to take from this work is that visual information can be used effectively as part of a speech filtering system. The research presented here successfully used visual information to deal with a very challenging speech environment. Recognition tests performed found that the use of visual information enabled this system to outperform audio-only approaches. The use of poor quality visual information was also tested, with even low quality visual data found to improve results.

3.2.3.4 Limitations

While this research demonstrates the benefit of a multimodal approach, it has some limitations. In these experiments, trained HMMs are used, and the assumption is made that the target speaker is also part of the training set. This means that experiments have not been attempted with completely novel data, only with a limited selection of sentences from a single corpus, and this limits the possible practical application of this work. Individual HMMs are used for each speaker used in the testing and training process. Also, this approach deals with enhancement and recognition in parallel, rather than having an enhancement stage followed by a separate recognition process.

3.2.4 Visually Derived Wiener Filtering

3.2.4.1 Summary of Work and Previous Papers

Almajai and Milner [45] have developed a multimodal speech enhancement system that makes use of visually derived Wiener filtering [46]. This approach builds on previous published work by the same authors. Firstly, [47] demonstrated a high degree of audiovisual correlation between the spectral output of speech and the shape of the mouth, and then built on this to filter speech by making use of visual features to estimate a corresponding noiseless audio signal, and then filtering a noisy audio signal [45, 48–50]. Wiener filtering works by comparing a noisy input signal to an estimation of an equivalent noiseless signal. Almajai and Milner [45] first created a basic Wiener filtering approach that initially made use of a simple joint audiovisual model and basic competing white noise, and then expanded upon it to produce a more sophisticated and comprehensive audiovisual speech enhancement system [49].

3.2.4.2 Key Output

Recent work from [49] uses sentences from an audiovisual speech database, spoken by a single male speaker. Visual information was recorded using a head mounted camera, and 2D-DCT was used to extract relevant lip information. This was then upsampled to match the equivalent audio information. In a system containing a noisy time domain audio signal $y(n)$ (with n representing sample number) and visual information taken from the facial region $v(i)$, with i representing frame number, $v(i)$ is used to produce an estimate of the log filterbank vector of the noiseless audio signal $\hat{x}(i)$. This is transformed into a linear filterbank estimate $L_{\hat{x}}(m)$, with m being a filterbank channel. This is then compared to an estimation of the noiseless speech plus noise. A noise only estimate, $L_y(m)$, is calculated from noise only periods of the utterance, identified with the aid of an audiovisual VAD. The combination of the speech and noise estimates is used to calculate the filterbank Wiener filter $L_w(m)$, which compares the noise-free estimate to the speech plus noise estimate given in Eq. (3.1),

$$L_w(m) = \frac{L_{\hat{x}}(m)}{L_{\hat{x}}(m) + L_y(m)} \tag{3.1}$$

The authors then interpolate $L_w(m)$ in order to match the dimensionality of the power spectrum of the audio signal to produce the frequency domain Wiener filter. This is used to calculate the enhanced speech power spectrum, which is then combined with the phase of the audio input and an inverse Fourier transform is used to return the enhanced speech to the time domain.

With Wiener filtering, the most complex aspect is the method of estimating the noise-free signal. A MAP estimate of the noise-free speech can be found with the use of visual information. The authors make use of phoneme-specific estimation. 36 monophone HMMs, plus an additional one for silence are trained using the training dataset. The training dataset makes use of forced Viterbi alignment to split each training utterance into phoneme sequences, and these labelled utterances are split into vector pools, with EM clustering used to train a GMM for each phoneme. The resulting HMMs are then used for speech estimation.

In the test sentences, an audiovisual speech recogniser is used to identify the phoneme. It is assumed in this work that the first few frames of the utterance are noise only, and an estimate of the SNR is taken from these. The speech recogniser then combines audio and visual recognition to identify the phoneme. This identifies the most suitable GMM to use for each frame, and returns the log filterbank speech-only noise-free estimate, $\hat{x}(i)$.

To estimate the noise-alone signal, an average of the non-speech vectors preceding the speech frame is taken. The SNR estimate, taken from the first few frames of the utterance (assumed to be noise only) is used to define how much weight to apply to the audio information. The frames identified as non speech are averaged, and this produces the noise alone estimate, $L_y(m)$. The results presented are generally positive, with both objective and subjective scores displaying the potential of this work.

3.2.4.3 Strengths

Fundamentally, this work demonstrates the potential for using visual information purely as part of a multimodal speech enhancement system, with the visual information used to remove noise from speech. This is a sophisticated system that takes account of the level of noise when it comes to phoneme decoding, and filters the signal differently depending on the phoneme identified. It can be seen that objective speech evaluation all showed a significant improvement in speech enhancement performance when compared to the original noisy speech and a standard audio-only spectral subtraction approach. Subjective human listening tests also showed that visually derived Wiener filtering was effective at removing noise from speech, with a significant improvement being found at all reported SNR test levels from 20 to 5 dB.

3.2.4.4 Limitations

There are a number of limitations. Firstly, when the results are analysed, the most significant results to consider are the subjective listening tests. While the work of the authors has reduced the noise intrusion score, this is at the cost of increasing the speech distortion score (see Chap. 5 for a full description of these metrics), meaning that even at relatively low SNR levels; listeners still have a slight preference for unfiltered speech over visually derived speech. Additionally, the results are limited to relatively high SNR levels (+5 to +20 dB), meaning that the system has not been tested in noisy environments. The model also makes use of a complex phoneme dependent model, and when forced alignment is used (i.e. manual labelling of phonemes), slightly improved results are obtained over standard automatic phoneme recognition. This is because in the presence of noise, phoneme decoding accuracy falls to 30 % at 0 dB, meaning that the accuracy of this system in noisy conditions is poor.

Another limitation with this work is the relatively constrained range of the database used for training. An audiovisual database containing 277 utterances for a single speaker is used, meaning it is effectively only trained and tested with input from the same speaker. The speech estimation model is also trained with training data from the same speaker, meaning that there is a potential lack of robustness in the GMMs used. Another issue is that the visual filtering approach makes use of visual information tracked by a camera, with the relevant ROI acquired with the aid of AcAMs; however, the system proposed in this research does not take account of situations where a poor visual feature-extraction result is returned.

When the Wiener filtering approach makes use of a VAD to determine speech and non speech, it then calculates a noise-only estimate. However, this estimate makes the assumption that the first few frames of the noisy speech signal are non-speech. The noisy speech itself is also described as "corrupted", but it is not specified whether the noise is simple additive noise, or whether a more complex mixing matrix is used to provide a more realistic speech filtering challenge. There is also the issue that speech is articulated differently in the presence of noise, as described by the well-known Lombard Effect [51, 52], which is not accounted for in the training of this system.

3.3 Visual Tracking and Detection

One aspect of relevance with regard to the work presented in this book is the extraction of relevant ROI information. This is an area of active research and development [53, 54]. It is felt to be relevant to summarise some recent developments in this research domain.

3.3.1 Lip Tracking

Lip tracking is an active field of research, with many different examples in the literature, such as Shape Models [55], and ACM [56]. Lip tracking represents a challenging research area, as it can be difficult to track lip images due to issues such as a weak colour contrast between skin and lip areas [57], and also the elastic shape and non rigid movement of the lips during speech [58]. A brief summary of a number of recent developments is discussed here.

Cheung et al. [58] divide lip tracking approaches into two main categories, edge based approaches, and region based approaches. Edge based approaches, as suggested by the name, rely on colour and edge information to track movement. This can rely on identifying colour contrasts [59, 60], key points [61], or points considered to be 'corners', as proposed by [57]. These approaches work well under desired conditions (i.e. a clean background and distinct features), but will produce poor results if the image is not ideal (for example, in poor lighting conditions or when the subject is wearing cosmetics that may interfere with contrast detection [58]. Some other approaches include the use of ACM [56], also known as 'Snakes', to detect edges.

ACM were first proposed in 1988 by [56] and are designed to fit lines (hence the reason they are known as 'snakes') to specific shapes for feature extraction. In the context of lip tracking, this means fitting a contour around the edge of the lips in order to identify and extract the lip shape. Snakes are based on minimisation of energy and operate by identifying edges. The contour is shaped by the idea of external and internal energy. Ideally, internal energy is minimised when the snake has a shape relevant to the desired object, and external energy is minimised when the snake has correctly identified the boundary of the desired object. There are many implementations of this technique for edge based lip tracking, and some examples of this approach include work by [56, 62].

Another example of an edge based approach is to use corner detection [57]. This technique converts images to binary images, identifies the lower half of a face, and then uses horizontal profile projection [63] (defined as the sum of pixel intensities in each row of an image) to identify the rows of an image corresponding to the lip region. This is done by identifying the two maximum points of the horizontal projection vector for the lower part of the face. Of this region, corners are identified, using the Harris Corner Detector proposed by [64].

Region based approaches primarily make use of region based information. Cheung et al. [58] identify four main categories of interest, Deformable Templates, Region based ACM approaches, AcAMs and ASMs. Firstly, ASMs [65, 66] use a set of landmark points, derived manually from a training set of images, to create a sample template to apply to the area of interest. When presented with a new image, the template is applied, and the points are then are iteratively moved to match the face (known as fitting). This is a relatively simplistic approach, although is relatively fast. It does require intensive and time consuming training however, and suffers from a lack of robustness with images not similar to those found in the training set.

AcAMs [67] operate by creating statistical models of visual features, making use of shape (as described above) and texture information. The shape is defined as geometrical information that remains when location, scale and rotational effects are filtered out from an object, and the texture refers to the pixel intensities across the object. The initial shape models are combined with grey-level variation in a single statistical appearance model. Models are trained with manually labelled test-sets, and can then be applied to unseen images. Using AcAMs has the advantage that detailed models can be produced, while still being relatively computationally efficient. The disadvantage with this technique is that the models require time consuming and intensive training before use, and can struggle to generalise accurately when presented with novel data.

Deformable Templates were introduced by [68], and operate in a similar way to the ACM approach described above. An initial template is specified, with parameters matching a lip shape. The minimisation of energy approach is then utilised to alter these parameters, and then as the parameters are adjusted, the template is altered to gradually match the boundary of the desired lip shape [54, 69]. The limitation of this approach is that if there is a very irregular lip shape, or the image has a very widely open mouth, then poor performance has been found. Region based ACM approaches, are similar to the ACM approaches defined above, but rather than searching for an edge, specific regions within an image are inspected in order to minimise energy by dividing images into lip and non lip regions [53, 58]. This approach is generally highly dependent on initial parameter initialisation.

3.3.2 Region of Interest Detection

While these represent some lip tracking options, as can be seen from the examples above, the initial lip detection and the lip tracking are often initialised in two distinct steps [55, 58], and are dependent on initial parameters being accurately defined. Therefore it is also of interest to consider overall ROI detection algorithms that can be used and adapted to automatically detect regions of interest and generate initial parameters.

Yang et al. [70] in a paper from 2002, and [71] classify face detection approaches into four key categories. Firstly, knowledge based approaches such as proposed in work by [72, 73] are rule based systems. These make use of rules, often defined by human experts with facial features represented by differences and positions from each other. Feature invariant approaches work by extracting structural features of the face, specifically focusing on identifying features that will be identifiable, even in environments when conditions like lighting and pose vary greatly. Some examples of this approach include [74, 75]. Template matching methods use standard patterns of a face which have been trained and stored, either for identifying entire faces, or for individual features. To detect ROI information, the correlations between the input image and stored patterns are computed for detection.

Appearance based methods are the final category of approaches identified by [70]. This category describes models that are learned from training images. Rather than templates, which are manually trained and configured, appearance based approaches rely purely on trained true or false results from the training data. This approach requires a considerable quantity of training data in order to be effective. Yang et al. identified a number of approaches used, such as Eigenfaces [76], Distribution-Based Methods [77], and Neural Networks [78]. This category of approaches is the most common approach used in state of the art research in this field, and specifically, the development of the Viola-Jones approach, pioneered by [36], is one of the most influential developments in recent years.

3.3.2.1 The Viola-Jones Detector

The Viola-Jones detector [36] is arguably one of the most important developments in the field of face detection. This is an appearance based method, with three main components, the integral image, classifier learning with adaboost, and an attentional cascade structure. The first aspect of the Viola-Jones detector to consider is the integral image, also known as the summed area table. This is a technique used for quickly computing the sum of values in a rectangular subset of a grid. This was first introduced to the field by [79], and is used for rapid calculation of Haar-Like features. Essentially, the value at any point of an integral image is the sum of all pixels above and to the left of that point. It can be computed efficiently, and then any rectangle in that integral image can be evaluated quickly. This is used in the Viola-Jones detector for the calculation of Haar-Like features.

Haar-Like features represent an improvement on calculating all image intensities. They were adapted from Haar wavelets, and use the integral image technique to calculate the sum of intensities in specific rectangular regions within an image of interest. The sum of intensities can then be compared for neighbouring regions, and then the difference between each of these regions can be calculated. The Viola-Jones detector uses comparisons of 2, 3, and 4 rectangles as part of the detection process. The detector requires a very intensive training process, requiring many hours of training images. Trained Haar-Like features are available as part of the OpenCV library [80], limiting the requirement for further training. There have been many proposed refinements to the original Haar-Like features, and a comprehensive description of these can be found in a detailed review by [81].

In order to identify the optimal features, an approach known as boosting [82] is utilised. This aims to produce a very accurate hypothesis of a classification result by combining many weak classifiers. The initial approach utilised by [36] is a modified version of the Adaboost [83] algorithm. The theory behind this approach is that the number of Haar-Like features in any image sub-window will naturally be very large, and in order to produce a usable and quick classification, the vast majority of features must be excluded, with focus given to a very small number of critical features. At each stage of the boosting process, a weak learning algorithm is designed to select one single Haar-Like feature that best separates two distinct regions with the minimum number of errors. Each weak classifier only depends on a single feature, and at each stage of the boosting process, the strongest weak classifiers are weighted accordingly to produce the overall classification with the least errors. With regard to face detection, [36] found in their testing that the first feature to be selected was a large feature demonstrating a strong difference between the eye region and the upper cheek, and the second feature was the contrast in image intensity between the two eye regions and the bridge of the nose. There have been a number of refinements to the original Adaboost algorithm, such as Gentleboost [84, 85], Realboost [86–88], and JS-Boost [89]. Again, a very detailed summary is given by [81].

The final component used in the Viola-Jones detector makes use of an attentional cascade structure. As has been stated previously, most of the many sub-windows produce a negative result, and so are not of relevance for classification. The cascade structure proposed by [36] aims to exploit this to reduce computation time, by using a tree (cascade) of trained classifiers. Simpler trained classifiers are used in the early stages to reject the majority of sub-windows, with more complex classifiers used in later stages. A negative result at any stage results in the sub-window being rejected. However, one drawback of this approach is that the training of classifiers at each stage was found to be extremely time consuming, with timescales of months talked about for early versions of face detectors.

3.4 Audiovisual Speech Corpora

There are a number of audiovisual corpora now available for use as part of an audiovisual system. While there are many audio-only speech corpora available for use, good quality and large-scale audiovisual speech databases are less widely available. Many such as CUAVE [90] are small scale corpora designed for specific tasks, with often only the collection of isolated words or digits. For more general speech processing use, larger databases with a range of speakers and sentences are more useful.

3.4.1 The BANCA Speech Database

The BANCA audiovisual speech database [91] was primarily designed for biometric authentication. It consists of a wide range of speech sentences (208) recorded from across Europe, with data recorded in four languages, and in a range of different scenarios. Two types of camera, good and poor quality were used, two different quality audio recordings were used, and data was recorded in three different environmental scenarios (controlled, degraded, adverse). Fifty two subjects were used in different scenarios. In each recording, the speaker was expected to provide two items of speech information. Both of these consisted of data associated with biometric authentication, i.e. a series of numbers, and a name, address, and date of birth.

3.4.2 The Extended M2VTS Database

XM2VTSDB [92] is designed for authentication and biometric purposes, with a large quantity of data to enable security-focused multimodal recognition system training, and contains data from 295 British-English speakers, with each speaker reading the same three sentences "0 1 2 3 4 5 6 7 8 9", "5 0 6 9 2 8 1 3 7 4", and "Joe took fathers green shoe bench out".

3.4.3 The AVICAR Speech Database

AVICAR [51] was designed for speech recognition. It contains video files and the associated audio files for 100 speakers from a range of backgrounds, and is recorded in a noisy environment. This database takes account of the Lombard Effect [51, 52] and accommodates a range of conditions by recording speakers in an automobile. This provides different levels of background noise (i.e. different car speeds, with associated varying levels of engine noise and wind effects), and there are four cameras recording the speaker from different angles, providing significant levels of raw data to work with. Each speaker recites a range of data, including single digits, single letters, phone numbers, and full speech sentences.

3.4.4 The VidTIMIT Multimodal Database

VidTIMIT is a non-invasive Australian-English audiovisual speech corpus, recorded by [93, 94]. It contains videos split into image sequences and matching audio files for 43 speakers reciting a number of phonetically balanced Timit sentences. There are ten sentences recorded for each speaker, with each subject speaking a variety of sentences. This corpus is useful, with a suitable range of speakers and sentences. However, the main disadvantage of this corpus is the presence of continuous background noise.

3.4.5 The GRID Corpus

The GRID corpus [44] is an English language multimodal speech corpus that contains 1000 sentences from each of 34 speakers, 18 male and 16 female. This corpus contains a number of videos of each speaker. Like VidTIMIT, there are no physical restrictions placed on the speaker. Additionally, the data is recorded in a visually and acoustically clean environment. Sentences were recorded in the format of "COMMAND COLOUR PREPOSITION LETTER DIGIT ADVERB", with each word being changeable, producing example sentences such as "Put red at G 9 now."

References

1. K. Chung, Challenges and recent developments in hearing aids. Part i. Speech understanding in noise, microphone technologies and noise reduction algorithms. Trends Amplif. **8**(3), 83–124 (2004)
2. T. Ricketts, H. Mueller, Making sense of directional microphone hearing aids. Am. J. Audiol. **8**(2), 117 (1999)
3. M. Valente, Use of microphone technology to improve user performance in noise, *Textbook of Hearing Aid Amplification* (Singular Thomason Learning, San Diego, 2000), p. 247
4. F. Kuk, D. Keenan, C. Lau, C. Ludvigsen, Performance of a fully adaptive directional microphone to signals presented from various azimuths. J. Am. Acad. Audiol. **16**(6), 333–347 (2005)
5. M. Cord, R. Surr, B. Walden, L. Olson, Performance of directional microphone hearing aids in everyday life. J. Am. Acad. Audiol. **13**(6), 295–307 (2002)
6. M. Cord, R. Surr, B. Walden, O. Dyrlund, Relationship between laboratory measures of directional advantage and everyday success with directional microphone hearing aids. J. Am. Acad. Audiol. **15**(5), 353–364 (2004)
7. T. Ricketts, P. Henry, Evaluation of an adaptive, directional-microphone hearing aid: evaluación de un auxiliar auditivo de micrófono direccional adaptable. Int. J. Audiol. **41**(2), 100–112 (2002)
8. R. Bentler, C. Palmer, A. Dittberner, Hearing-in-noise: comparison of listeners with normal and (aided) impaired hearing. J. Am. Acad. Audiol. **15**(3), 216–225 (2004)
9. L. Mens, Speech understanding in noise with an eyeglass hearing aid: asymmetric fitting and the head shadow benefit of anterior microphones. Int. J. Audiol. **50**(1), 27–33 (2011)
10. L. Christensen, D. Helmink, W. Soede, M. Killion, Complaints about hearing in noise: a new answer. Hear. Rev. **9**(6), 34–36 (2002)
11. S. Laugesen, T. Schmidtke, Improving on the speech-in-noise problem with wireless array technology. News from Oticon (2004), pp. 3–23
12. S. Rosen, Temporal information in speech: acoustic, auditory and linguistic aspects. Philos. Trans.: Biol. Sci. **336**, 367–373 (1992)
13. N. Tellier, H. Arndt, H. Luo, Speech or noise? Using signal detection and noise reduction. Hear. Rev. **10**(6), 48–51 (2003)
14. H. Levitt, Noise reduction in hearing aids: an overview. J. Rehabil. Res. Dev. **38**(1), 111–121 (2001)
15. M. Boymans, W. Dreschler, P. Schoneveld, H. Verschuure, Clinical evaluation of a full-digital in-the-ear hearing instrument. Int. J. Audiol. **38**(2), 99–108 (1999)
16. J. Alcántara, B. Moore, V. Kühnel, S. Launer, Evaluation of the noise reduction system in a commrcial digital hearing aid: evaluación del sistema de reducción de ruido en un auxiliar auditivo digital comercial. Int. J. Audiol. **42**(1), 34–42 (2003)
17. C. Elberling, About the voicefinder. News from Oticon (2002)

18. D. Schum, Noise-reduction circuitry in hearing aids: (2) goals and current strategies. Hear. J. **56**(6), 32 (2003)
19. L. Girin, J. Schwartz, G. Feng, Audio-visual enhancement of speech in noise. J. Acoust. Soc. Am. **109**, 3007 (2001)
20. R. Goecke, G. Potamianos, C. Neti, Noisy audio feature enhancement using audio-visual speech data, in *Proceedings of the IEEE International Conference on Acoustics, Speech, and Signal Processing (ICASSP'02)*, vol. 2 (IEEE, 2002), pp. 2025–2028
21. S. Deligne, G. Potamianos, C. Neti, Audio-visual speech enhancement with AVCDCN (audio-visual codebook dependent cepstral normalization), in *Proceedings of the Sensor Array and Multichannel Signal Processing Workshop* (IEEE, 2003), pp. 68–71
22. A. Acero, R. Stern, Environmental robustness in automatic speech recognition, in *Proceedings of the International Conference on Acoustics, Speech, and Signal ProcessingICASSP-90* (IEEE, 2002), pp. 849–852
23. L. Deng, A. Acero, L. Jiang, J. Droppo, X. Huang, High-performance robust speech recognition using stereo training data, in *Proceedings of the IEEE International Conference on Acoustics, Speech, and Signal Processing (ICASSP'01)*, vol. 1 (IEEE, 2002), pp. 301–304
24. B. Rivet, J. Chambers, Multimodal speech separation, in *Advances in Nonlinear Speech Processing*, vol. 5933, Lecture Notes in Computer Science, ed. by J. Sole-Casals, V. Zaiats (Springer, Berlin, 2010), pp. 1–11
25. B. Rivet, L. Girin, C. Jutten, Log-Rayleigh distribution: a simple and efficient statistical representation of log-spectral coefficients. IEEE Trans. Audio Speech Lang. Process. **15**(3), 796–802 (2007)
26. B. Rivet, L. Girin, C. Jutten, Mixing audiovisual speech processing and blind source separation for the extraction of speech signals from convolutive mixtures. IEEE Trans. Audio Speech Lang. Process. **15**(1), 96–108 (2007)
27. B. Rivet, L. Girin, C. Serviere, D.-T. Pham, C. Jutten, Using a visual voice activity detector to regularize the permutations in blind separation of convolutive speech mixtures, in *Proceedings of the 15th International Conference on Digital Signal Processing* (2007), pp. 223 –226
28. B. Rivet, L. Girin, C. Jutten, Visual voice activity detection as a help for speech source separation from convolutive mixtures. Speech Commun. **49**(7–8), 667–677 (2007)
29. C. Jutten, J. Herault, Blind separation of sources, part I: an adaptive algorithm based on neuromimetic architecture. Signal Process. **24**(1), 1–10 (1991)
30. J. Herault, C. Jutten, B. Ans, Detection de grandeurs primitives dans un message composite par une architecture de calcul neuromimetrique en apprentissage non supervise. Actes du Xeme colloque GRETSI **2**, 1017–1020 (1985)
31. E. Cherry, Some experiments on the recognition of speech, with one and with two ears. J. Acoust. Soc. Am. **25**(5), 975–979 (1953)
32. L. Girin, G. Feng, J. Schwartz, Fusion of auditory and visual information for noisy speech enhancement: a preliminary study of vowel transitions, in *Proceedings of the 1998 IEEE International Conference on Acoustics, Speech and Signal Processing*, vol. 2 (IEEE, 2002), pp. 1005–1008
33. D. Sodoyer, L. Girin, C. Jutten, J. Schwartz, Developing an audio-visual speech source separation algorithm. Speech Commun. **44**(1–4), 113–125 (2004)
34. D. Sodoyer, J. Schwartz, L. Girin, J. Klinkisch, C. Jutten, Separation of audio-visual speech sources: a new approach exploiting the audio-visual coherence of speech stimuli. EURASIP J. Appl. Signal Process. **2002**(1), 1165–1173 (2002)
35. S. Naqvi, M. Yu, J. Chambers, A multimodal approach to blind source separation of moving sources. IEEE J. Sel. Top. Signal Process. **4**(5), 895–910 (2010)
36. P. Viola, M. Jones, Rapid object detection using a boosted cascade of simple features, in *Proceedings of the IEEE Computer Society Conference on Computer Vision and Pattern Recognition*, vol. 1 (IEEE Computer Society, 2001), pp. 511–518
37. A. Hyvarinen, J. Karhunen, E. Oja, *Independent Component Analysis*, vol. 26 (Wiley-Interscience, New York, 2001)

38. E. Bingham, A. Hyvarinen, A fast fixed-point algorithm for independent component analysis of complex valued signals. Int. J. Neural Syst. **10**(1), 1–8 (2000)
39. J. Barker, X. Shao, Audio-visual speech fragment decoding, in *Proceedings of the International Conference on Auditory-Visual Speech Processing* (2007), pp. 37–42
40. J. Barker, M. Cooke, D. Ellis, Decoding speech in the presence of other sources. Speech Commun. **45**(1), 5–25 (2005)
41. A. Bregman, *Auditory Scene Analysis: The Perceptual Organization of Sound* (The MIT Press, Cambridge, 1990)
42. A. Bregman, *Auditory Scene Analysis: Hearing in Complex Environments* (Oxford University Press, Oxford, 1993)
43. J. Barker, X. Shao, Energetic and informational masking effects in an audiovisual speech recognition system. IEEE Trans. Audio Speech Lang. Process. **17**(3), 446–458 (2009)
44. M. Cooke, J. Barker, S. Cunningham, X. Shao, An audio-visual corpus for speech perception and automatic speech recognition. J. Acoust. Soc. Am. **120**(5 Pt 1), 2421–2424 (2006)
45. I. Almajai, B. Milner, in *Proceedings of the Enhancing Audio Speech using Visual Speech Features* (Interspeech, Brighton, 2009)
46. N. Wiener, *Extrapolation, Interpolation, and Smoothing of Stationary Time Series: With Engineering Applications* (The MIT Press, Cambridge, 1949)
47. I. Almajai, B. Milner, Maximising audio-visual speech correlation, in *Proceedings of the AVSP* (2007)
48. I. Almajai, B. Milner, J. Darch, S. Vaseghi, Visually-derived Wiener filters for speech enhancement, in *Proceedings of the IEEE International Conference on Acoustics, Speech and Signal Processing, ICASSP*, vol. 4 (2007), pp. 585–588
49. I. Almajai, B. Milner, in *Proceedings of the Effective Visually-derived Wiener Filtering For Audio-visual Speech Processing* (Interspeech, Brighton, UK, 2009)
50. B. Milner, I. Almajai, Noisy audio speech enhancement using Wiener filters derived from visual speech, in *Proceedings of the International Workshop on Auditory-Visual Speech Processing (AVSP)*
51. B. Lee, M. Hasegawa-Johnson, C. Goudeseune, S. Kamdar, S. Borys, M. Liu, T. Huang, AVICAR: audio-visual speech corpus in a car environment, in *Proceedings of the Conference on Spoken Language, Jeju, Korea* (Citeseer, 2004), pp. 2489–2492
52. H. Lane, B. Tranel, The Lombard sign and the role of hearing in speech. J. Speech Hear. Res. **14**(4), 677 (1971)
53. T. Wakasugi, M. Nishiura, K. Fukui, Robust lip contour extraction using separability of multi-dimensional distributions, in *Proceedings of the Sixth IEEE International Conference on Automatic Face and Gesture Recognition* (IEEE, 2004), pp. 415–420
54. A. Liew, S. Leung, W. Lau, Lip contour extraction from color images using a deformable model. Pattern Recognit. **35**(12), 2949–2962 (2002)
55. Q. Nguyen, M. Milgram, Semi adaptive appearance models for lip tracking, in *Proceedings of the ICIP09* (2009), pp. 2437–2440
56. M. Kass, A. Witkin, D. Terzopoulos, Snakes: active contour models. Int. J. Comput. Vis. **1**, 321–331 (1988)
57. A. Das, D. Ghoshal, Extraction of time invariant lips based on morphological operation and corner detection method. Int. J. Comput. Appl. **48**(21), 7–11 (2012)
58. Y. Cheung, X. Liu, X. You, A local region based approach to lip tracking. Pattern Recognit. **45**, 3336–3347 (2012)
59. X. Zhang, R. Mersereau, Lip feature extraction towards an automatic speechreading system, in *Proceedings of the 2000 International Conference on Image Processing*, vol. 3 (IEEE, 2000), pp. 226–229
60. N. Eveno, A. Caplier, P. Coulon, New color transformation for lips segmentation, in *IEEE Fourth Workshop on Multimedia Signal Processing* (IEEE, 2001), pp. 3–8
61. N. Eveno, A. Caplier, P. Coulon, Key points based segmentation of lips, in *Proceedings of the 2002 IEEE International Conference on Multimedia and Expo, ICME'02*, vol. 2, (IEEE, 2002), pp. 125–128

62. D. Freedman, M. Brandstein, Contour tracking in clutter: a subset approach. Int. J. Comput. Vis. **38**(2), 173–186 (2000)
63. Z. Ji, Y. Su, J. Wang, R. Hua, Robust sea-sky-line detection based on horizontal projection and hough transformation, in *2nd International Congress on Image and Signal Processing, CISP'09* (IEEE, 2009), pp. 1–4
64. C. Harris, M. Stephens, A combined corner and edge detector, in *Alvey Vision Conference*, vol. 15 (Manchester, 1988), p. 50
65. J. Luettin, N. Thacker, S. Beet, Visual speech recognition using active shape models and hidden Markov models, in *Proceedings of the IEEE International Conference on Acoustics, Speech, and Signal Processing, ICASSP-96*, vol. 2 (IEEE, 1996), pp. 817–820
66. Q. Nguyen, M. Milgram, T. Nguyen, Multi features models for robust lip tracking, in *10th International Conference on Control, Automation, Robotics and Vision, 2008. ICARCV 2008*, (IEEE, 2008), pp. 1333–1337
67. T. Cootes, G. Edwards, C. Taylor, Active appearance models, in*Computer Vision-ECCV'98* (1998), pp. 484–498
68. A. Yuille, P. Hallinan, D. Cohen, Feature extraction from faces using deformable templates. Int. J Comput. Vis. **8**(2), 99–111 (1992)
69. G. Chiou, J. Hwang, Lipreading from color video. IEEE Trans. Image Process. **6**(8), 1192–1195 (1997)
70. M. Yang, D. Kriegman, N. Ahuja, Detecting faces in images: a survey. IEEE Trans. Pattern Anal. Mach. Intell. **24**(1), 34–58 (2002)
71. S. Wang, A. Abdel-Dayem, Improved viola-jones face detector, in *Proceedings of the 1st Taibah University International Conference on Computing and Information Technology, ICCIT'12* (2012), pp. 321–328
72. C. Kotropoulos, I. Pitas, Rule-based face detection in frontal views, in *IEEE International Conference on Acoustics, Speech, and Signal Processing, ICASSP-97*, vol. 4, (IEEE, 1997), pp. 2537–2540
73. G. Yang, T. Huang, Human face detection in a complex background. Pattern Recognit. **27**(1), 53–63 (1994)
74. R. Kjeldsen, J. Kender, Finding skin in color images, in *Proceedings of the Second International Conference on Automatic Face and Gesture Recognition* (IEEE, 1996), pp. 312–317
75. K. Yow, R. Cipolla, A probabilistic framework for perceptual grouping of features for human face detection, in *Proceedings of the Second International Conference on Automatic Face and Gesture Recognition*, (IEEE, 1996), pp. 16–21
76. T. Kohonen, *Self-organisation and Associative Memory* (Springer, Berlin, 1989)
77. K. Sung, Learning and example selection for object and pattern detection (1996)
78. T. Agui, Y. Kokubo, H. Nagahashi, T. Nagao, Extraction of face regions from monochromatic photographs using neural networks, in *Proceedings of the International Conference on Robotics* (1992)
79. F. Crow, Summed-area tables for texture mapping. Comput. Graph. **18**(3), 207–212 (1984)
80. G. Bradski, The OpenCV Library. Dr. Dobb's J. Softw. Tools **25**(11), 120–126 (2000)
81. C. Zhang, Z. Zhang, A survey of recent advances in face detection. *Microsoft Research*, June 2010
82. R. Meir, G. Rätsch, An introduction to boosting and leveraging, *Advanced Lectures on Machine Learning* (Springer, New York, 2003), pp. 118–183
83. Y. Freund, R. Schapire, A decision-theoretic generalization of on-line learning and an application to boosting, *Computational Learning Theory* (Springer, Berlin, 1995), pp. 23–37
84. J. Friedman, T. Hastie, R. Tibshirani, Additive logistic regression: a statistical view of boosting (with discussion and a rejoinder by the authors). The Ann. Stat. **28**(2), 337–407 (2000)
85. S. Brubaker, J. Wu, J. Sun, M. Mullin, J. Rehg, On the design of cascades of boosted ensembles for face detection. Int. J. Comput. Vis. **77**(1), 65–86 (2008)
86. S. Li, L. Zhu, Z. Zhang, A. Blake, H. Zhang, H. Shum, Statistical learning of multi-view face detection. in *Computer Vision, ECCV 2002* (2006), pp. 117–121

87. C. Bishop, P. Viola, Learning and vision: discriminative methods. ICCV Course Lear. Vis. **2**(7), 11 (2003)
88. R. Schapire, Y. Singer, Improved boosting algorithms using confidence-rated predictions. Mach. Lear. **37**(3), 297–336 (1999)
89. X. Huang, S. Li, Y. Wang, Jensen-Shannon boosting learning for object recognition, in *IEEE Computer Society Conference on Computer Vision and Pattern Recognition, CVPR*, vol. 2 (IEEE, 2005), pp. 144–149
90. E. Patterson, S. Gurbuz, Z. Tufekci, J. Gowdy, Cuave: a new audio-visual database for multimodal human-computer interface research, in *IEEE International Conference on Acoustics, Speech, and Signal Processing, ICASSP-93*, vol. 2 (IEEE, 2002), p. II
91. E. Bailly-Bailliere, S. Bengio, F. Bimbot, M. Hamouz, J. Kittler, J. Mariéthoz, J. Matas, K. Messer, V. Popovici, F. Porée et al., The BANCA database and evaluation protocol, *Audio- and Video-Based Biometric Person Authentication* (Springer, 2003), p. 1057
92. K. Messer, J. Matas, J. Kittler, J. Luettin, G. Maitre, XM2VTSDB: the extended M2VTS database, in *Second International Conference on Audio and Video-based Biometric Person Authentication*, vol. 964 (Citeseer, 1999), pp. 965–966
93. C. Sanderson, K. Paliwal, Polynomial features for robust face authentication, in *Proceedings of the International Conference on Image Processing*, vol. 3 (IEEE, 2002), pp. 997–1000
94. C. Sanderson, *Biometric Person Recognition: Face, Speech and Fusion* (VDM Verlag Dr, Muller, 2008)

Chapter 4
A Two Stage Multimodal Speech Enhancement System

Abstract The overall aim of this work is to utilise the relationship between audio and visual aspects of speech in order to develop a speech enhancement system. This chapter provides a detailed description of the initial two-stage multimodal speech enhancement system presented here. This represents a combination of a variety of state-of-the-art techniques, all integrated into one novel system. Each individual component is described in detail, covering feature extraction, audiovisual Wiener filtering, the audiovisual model required by this filtering approach, and audio-only beamforming. This system is described in this chapter and an evaluation of the strengths and weaknesses of this approach is presented in the following chapter. This chapter presents the technical description of this multimodal speech enhancement system.

Keywords Integrated system · Lip tracking · Speech enhancement · Multimodal · Filtering · Beamforming · Wiener filtering

4.1 Overall Design Framework of the Two-Stage Multimodal System

The overall aim of this work is to utilise the relationship between audio and visual aspects of speech in order to develop a speech enhancement system. The speech filtering system presented in this chapter is an extension of existing audio-only concepts, in that it extends the concept of an audio-only two-stage filtering system that combines multiple audio-only filtering techniques into one integrated system, as demonstrated in examples by [1, 2]. These systems are theoretically more powerful than those using only a single technique due to the additional filtering offered by utilising a combination of techniques and the addition of visual information may add more potential still. The system presented in this chapter extends this idea by combining audio beamforming with visually derived Wiener filtering to produce a novel integrated two-stage speech enhancement system, theoretically capable of functioning in extremely noisy environments. The overall diagram of this multimodal system is shown in Fig. 4.1.

A. Abel and A. Hussain, *Cognitively Inspired Audiovisual Speech Filtering*,
SpringerBriefs in Cognitive Computation, DOI 10.1007/978-3-319-13509-0_4

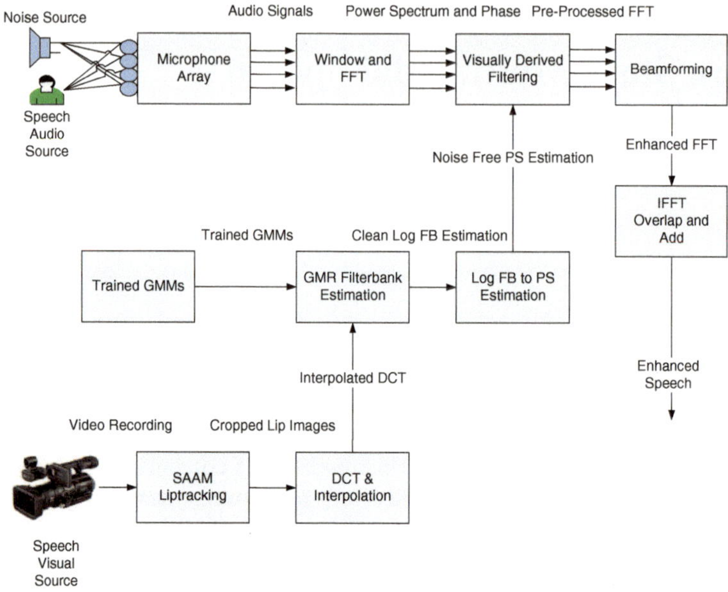

Fig. 4.1 Block diagram of multimodal two-stage filtering system components

The overall system diagram, as seen in Fig. 4.1, shows that the system is presented with two inputs. Firstly, there is the audio input, which consists of a mixed speech and noise source, and there is also a visual input, in the form of a video recording of a matching speech source. To summarise the components utilised in this system and shown in Fig. 4.1, the audio signal is received by the microphone array, and this signal is then windowed and transformed into the frequency domain. This audio signal is then used as the noisy input into a visually derived Wiener filtering process, using visual information to produce an estimate of the noise-free audio signal. This is then used to perform Wiener filtering. After this, the pre-processed signals are then filtered using audio-only beamforming to produce an enhanced frequency-domain signal. Finally, this is then transformed back to the time domain and output.

4.2 Reverberant Room Environment

In order for speech filtering to be performed in an experimental environment, the speech and noise sources have to be mixed. There are two main alternatives, additive or convolved mixtures. Additive mixtures are the most simple method of combining speech and noise, and simply consist of combining speech plus noise "speech+noise" to create a noisy speech mixture. Although this is the simplest type of mixture, a simple additive mixture does not take into account factors such as the difference in location of source and noise, atmospheric conditions such as temperature and humidity, or reverberation (a natural consequence of broadcasting sound in a room). Reverberation, in the context of this research, refers to the situation where large numbers of echoes are built up during the transmission of a sound due to environmental factors such as a small room. These echoes take time to dissipate, and so have an effect on the input received by a microphone or human listener. Convolved mixtures of speech and noise [3, 4] can provide a more realistic noisy speech mixture. These convolved mixtures do not simply add the sounds together, but necessitate the construction of a mixing matrix.

In this work, the noisy speech mixtures used are mixed in a convolved manner. To do this, a simulated room environment is used, with the speech and noise sources transformed with the matching impulse responses. Impulse responses represent the characteristics of a room when presented with a brief audio sample, and these are then applied to the speech and noise signals in the context of their location within the simulated room. This gives them the characteristics of being affected by environmental conditions with regard to microphone input. These sources are then convolved.

In order to create this speech mixture, the simulated room used in this work has a number of parameters that have to be defined. Firstly, it is assumed that the speech and noise sources originate at different locations within this simulated room. This room has been designed with dimensions of 5 by 4 by 3 m and this room is considered to be a closed room. It is assumed for the purposes of calculating speed of sound that the air temperature is 20 °C, with humidity of 40 %. The speech source is located at xyz coordinates of (2.50, 2.00, 1.40 m), and the noise source at (3.00, 3.00, 1.40 m). Finally, the simulated microphone input array is located at coordinates (2.20 m, y coordinates ranging from 1.88 to 2.12, 1m).

4.3 Multiple Microphone Array

As discussed, a multiple microphone array is used to receive the noisy speech mixtures. The reason for these multiple microphones is to allow the directional beamforming aspect of the integrated two-stage system to function. As shown in Fig. 4.1, in the work presented in this chapter, the noise and speech mixture is received by an array of four microphones, with each microphone at the same x coordinate, the same z coordinate, but slightly different y coordinates. The first microphone is positioned 2.20 m along the length of the room (x coordinate), at a height of 1 m (z coordinate), and with a y coordinate of 1.88 m. The other microphone are positioned at the same x and y coordinates, but with y coordinates of 1.96, 2.04, and 2.12 m respectively.

In the simulated room environment, it is assumed in this work that the noise and signal come from different sources. In this research, the noisy speech mixtures are received by an array of four microphones. To summarise this concept, assume M microphone signals, $z_1(t), \ldots, z_M(t)$ record a source $x(t)$ and M uncorrelated noise interfering signals $\hat{x}_1(t), \ldots, \hat{x}_M(t)$. Thus, the mth microphone signal is given by,

$$Z_m(t) = a_m(t) * x(t) + \hat{x}_m(t), \quad 1 \leq m \leq M \tag{4.1}$$

where $a_m(t)$ is the impulse response of the mth sensor to the desired source, and $*$ denotes convolution. In the frequency-domain convolutions become multiplications. Furthermore, since in this work there is no interest in balancing the channels, the source is redefined so that the first channel becomes unity. Hence, applying the Short-Time Fourier Transform (STFT) to (4.1), this results in,

$$Z_m(k,l) = A_m(k,l)X(k,l) + \hat{X}_m(k,l), \quad 1 \leq m \leq M \tag{4.2}$$

where k is the frequency bin index, and l the time-frame index. Thus, this produces a set of M equations that can be written in a compact matrix form as,

$$\mathbf{Z}(k,l) = \mathbf{A}(k,l)X(k,l) + \hat{\mathbf{X}}(k,l) \tag{4.3}$$

with,

$$\begin{aligned} \mathbf{Z}(k,l) &= [Z_1(k,l)Z_2(k,l) \ldots Z_M(k,l)]^T \\ \mathbf{A}(k,l) &= [A_1(k,l)A_2(k,l) \ldots A_M(k,l)]^T \\ \hat{\mathbf{X}}(k,l) &= [\hat{X}_1(k,l)\hat{X}_2(k,l) \ldots \hat{X}_M(k,l)]^T \end{aligned} \tag{4.4}$$

This produces the convolved Fourier transformed microphone mixtures of speech and noise, represented by $Z_1(k,l), \ldots, Z_4(k,l)$. Each of these represents the transformed noisy input of an individual microphone, sampled at 8 kHz. These input signals are then used to extract audio log filterbank features, described in Sect. 4.4, to produce the audio log filterbank values $F_a(1), \ldots, F_a(4)$ for each microphone input, using an audio frame of 25 ms with a 50 % overlap. At the same time, matching visual DCT features F_v are extracted from the video recordings, which will be described in Sect. 4.5.

4.4 Audio Feature Extraction

The previous section described the process of receiving noisy speech signals from the microphone inputs and transforming them to the frequency domain. These transformed signals are used in the second stage of this integrated multimodal system to carry out audio-only beamforming. Before the second filtering stage, the initial stage of the filtering process makes use of visually derived Wiener filtering. For this, the audio input has to be transformed to make it possible for audio to be estimated given visual data (as is described in more detail in Sect. 4.6). The initial Fourier transformed noise mixtures are further transformed to produce the magnitude spectrum, and subsequently log filterbank values. These are used as part of the visually derived filtering process.

Therefore, in the system presented in this chapter, a filterbank dimension of $M = 23$ filters is used. This is motivated by other work in the literature such as [5], which uses the same size of filterbank. The relationship between linear and Mel frequency is given by,

$$\hat{f}_{mel} = 2595 \cdot log10\left(1 + \frac{f_{lin}}{700}\right) \tag{4.5}$$

In this implementation, the limits of the frequency range are the parameters that define the basis for the filterbank design. The unit interval $\Delta\hat{f}$ is determined by the lower and the higher boundaries of the frequency range of the entire filter bank, $\hat{f}_{high}$ and $\hat{f}_{low}$, as follows,

$$\Delta\hat{f} = \frac{\hat{f}_{high} - \hat{f}_{low}}{M + 1} \tag{4.6}$$

the centre frequency $\hat{f}_{cm}$ of the mth filter is given by,

$$\hat{f}_{cm} = \hat{f}_{low} + m\Delta\hat{f}, \quad m = 1, \ldots, M - 1 \tag{4.7}$$

where M represents the total number of filters in the filterbank. The conversion of the centre frequencies of the filters to linear frequency (Hz) is given by,

$$\hat{f}_{cm} = 700 \cdot \left(10^{\hat{f}_{cm}/2595} - 1\right) \tag{4.8}$$

and the shape of the mth triangular filter is defined by,

$$H_m(k) = \begin{cases} 0 & k < f_{b_{m-1}} \\ \frac{k - f_{b_{m-1}}}{f_{b_m} - f_{b_{m-1}}} & f_{b_{m-1}} \leq k \leq f_{b_m} \\ \frac{f_{b_{m+1}} - k}{f_{b_{m+1}} - f_{b_m}} & f_{b_m} \leq k \leq f_{b_{m+1}} \\ 0 & k > f_{b_{m+1}} \end{cases} \quad m = 1, \ldots, M \tag{4.9}$$

where f_{b_m} are the boundary points of the filters and $k = 1, \ldots, K$ corresponds to the kth coefficient of the K-points Discrete Fourier Transform (DFT). The boundary points f_{b_m} are expressed in terms of position, which depends on the sampling frequency F_{samp} and the number of points K in the DFT,

$$f_{b_m} = \left(\frac{K}{F_s} \cdot f_{cm}\right) \tag{4.10}$$

For computing the log filterbank parameters, the magnitude spectrum $|Z(k)|$ of each mixed noisy audio signal acts as the input for the filterbank $H_m(k)$. Next, the filterbank output is logarithmically compressed to produce the log filterbank audio signal for further processing,

$$F_a = ln\left(\sum_{k=0}^{K-1} |Z(k)| \cdot H_m(k)\right) \tag{4.11}$$

This produces the log filterbank signal F_a of each microphone input, which is used, along with matching visual information (the extraction of which is discussed in the next section) to carry out visually derived speech filtering.

4.5 Visual Feature Extraction

In addition to the audio features discussed in the previous sections, in order to perform visually derived filtering, there is an obvious requirement for visual feature extraction. The visual filtering algorithm relies upon DCT vectors taken from lip information, and this section discusses the extraction of these features. Visual lip features are extracted by using the state-of-the-art Semi Adaptive Appearance Modules (SAAM) approach pioneered by [6] and also used in work carried out by the author in [7]. There are many different approaches to lip tracking and visual tracking in general, with much active research and many state of art commercial solutions such as the Microsoft Kinect, that makes use of infrared and Red Blue Green (RGB) cameras to track skeletal frames.

Extracting lip information by manually cropping each frame is extremely time consuming. The chosen automated approach had one key advantage in that it was considered state-of-the-art, and collaborative work took place with the developers of this technique [6] who also reported good results with this technique, to adapt and test this system in the context of multimodal speech processing, resulting in the publication of the first utilisation of the SAAM approach in the context of speech correlation in [7]. Another benefit of using this approach was that the chosen tracking technique was a standalone component that could be integrated into the system without difficulty. Its loose integration with the other components of this system also means that it is feasible for this approach to be replaced in future work with a different front end without any difficulty. The results of informally testing this SAAM approach with a range of different speech sentences from the GRID corpus found that the system was able to reliably and successfully track data from this corpus; making it suitable for use in this work.

There are three main components needed for the successful utilisation of visual information as part of a speech processing system. The first is accurate ROI detection. The second is the ability to automatically track and extract the ROI in each frame of a speech sentence, taking into consideration that the speaker may not remain completely still throughout. Finally, each cropped ROI frame has to be extracted and transformed into a format suitable for further processing.

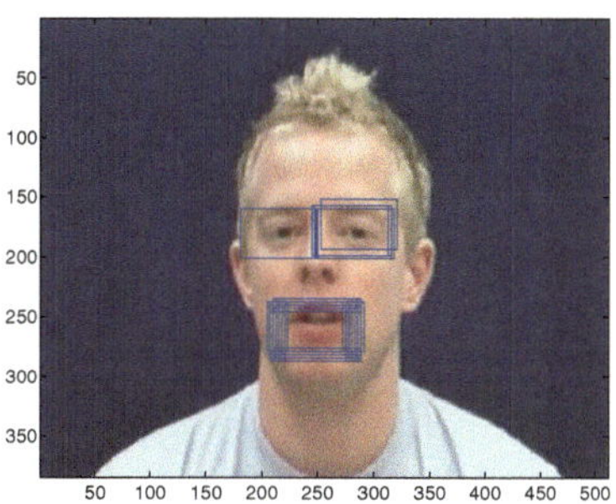

Fig. 4.2 Example of initial results from Viola-Jones Lip Detection process. *Rectangles* indicate a possible candidate ROI

While there are many lip tracking approaches that have been proposed in the literature [6, 8], some of these require the initial identification of the ROI, with approaches such as AcAMs [9] being highly dependent on the setting of initial parameters. Therefore, while the SAAM lip tracking approach is used in this work, the extraction of the ROI is a separate issue. As stated, there are many examples of both lip tracking and ROI detection [10, 11], and a summary of these was presented in Chap. 3. It was decided to make use of an implementation of the Viola-Jones [12] detector by [13], as described in Chap. 3.

The training of an appearance based lip detector can be a very intensive and time consuming process, and so in this work, it was decided to make use of Haar-Like features that have been specifically trained for lip detection and are available as part of the OpenCV [14] library, limiting the need for further training. The chosen lip detection implementation runs the algorithm on a single image frame, and then returns a number of potential candidate ROI areas. An example of this initial output is shown in Fig. 4.2.

It can be seen in the example given in Fig. 4.2 that a number of these candidates are correct, with the majority centred around the mouth region as would be expected. However, it can also be seen that a number of other potential matches are found, in the example of Fig. 4.2 a number of candidates can be seen around the left eye region of the image, as well as another candidate around the right eye. This is a common occurrence, and the final decision was made by customising the initial code in a similar manner to other work, including by [12]. The candidate areas were divided into a number of different subsets by firstly identifying overlapping candidates. If one candidate had an overlap with another of 65 % or greater (value chosen manually after investigation of preliminary test data), then the candidates were considered to belong to that subset. This had the effect of both dividing the candidates into subsets, and also removing values with only a small overlap. It was then assumed for the purposes of this research that the subset with the greatest number of candidate members was the relevant mouth area. This is because the data used for this research concentrates on single face scenarios.

To calculate the final ROI area, the mean of all the values within the most relevant subset was calculated, producing a single rectangular ROI. Testing (which will be discussed in more depth in Chap. 5) of the detector identified a number of issues.

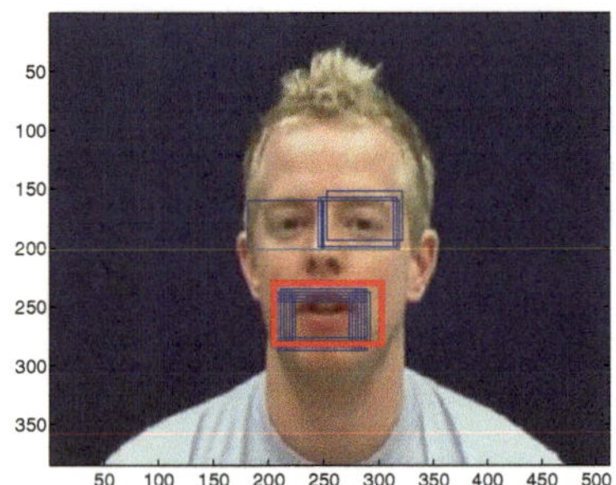

Fig. 4.3 Example of initial results from Viola-Jones Lip Detection process. *Thinner rectangles* indicate a possible candidate ROI, and the *thicker rectangle* represents the final selected mouth region to be used for tracking.

Firstly, to produce a more relevant ROI, it was found that the detected areas were smaller than ideal. This was solved by adding 20 pixels to the width and 10 to the height in order to contain more useful data. An example of this refined output can be seen in Fig. 4.3. This is the same image as Fig. 4.2, but the thicker rectangle represents the final output.

The second issue was that in a very limited number of cases, the detector did not successfully identify the lip region. The problem lies with the trained Haar-Like features, rather than in the code, and this was solved by running a check on the initial number of candidates. If an image was found to produce <7 candidate ROI areas, then a second scan of the image was run, using different Haar-Like features. These were trained to identify the whole face, and were found to work reliably in all tested incidences. The face region was then cropped to produce an estimate of the lip region. This lip detection was used in the first frame of the image to identify the ROI. As previously stated, the second aspect, lip tracking, is handled by using the SAAM technique. Finally, the extraction of visual information into a usable format (2D-DCT) is covered later in this section.

Lip tracking essentially deals with non-stationary data, as the appearance of a target object may alter drastically over time due to factors like pose, variation, and illumination changes. The lip tracking framework discussed is based on the Adaptive Appearance Models (AAM) approach by [15], which allows for the updating of the mean and Eigen vectors of g-dimensional observation vectors $v_o \in R^g$. First, the AAM technique is extended by inserting a supervisor model [16] that verifies the AAM performance at each frame in the sequence, by using a Support Vector Machine (SVM) to filter the AAM result for an individual frame, as shown in (4.12),

$$y\left(v_o\right) = sgn\left(\sum_i \alpha_i \hat{v}_i \omega\left(v_o i, v_o\right) + b\right) \tag{4.12}$$

Where $y(v_o) \in \{-1, 1\}$ signifies whether v_o represents a good or bad result. α, b are trained offline with the SVM [17], ω (.) is the Gaussian kernel function and v_{oi}, v are trained and observation vectors respectively. Each $\hat{v}_i$ represents the desired output of each example v_{Ti} from the offline training dataset.

Secondly, shape models are constructed to allow the SAAM technique to track feature points in video sequences. To model deformation, a shape model is formed,

$$S^{\circ} = S^{\circ} + P_s b \tag{4.13}$$

where $S^{\circ} = \left(v^{\circ}_{o1}, \hat{v}^{\circ}_{1}, \ldots, v^{\circ}_{oj}, \hat{v}^{\circ}_{j}\right)$ is a normalised shape and j represents a number of feature points. To track these, it is sufficient to find the parameters,

$$p = \left[b_{o1}, \ldots, \phi_v, \phi_q, \theta, s\right] \tag{4.14}$$

where b_i is the coefficient to deform S°, and ϕ_v, ϕ_q, θ, s represent translations, rotation and scale parameters respectively. To track a target object, the aim is to maximize the cost function given in (4.15) as follows,

$$p^{*} = \arg\max_{p}(y_e) \tag{4.15}$$

Where y_e is a negative exponential of projection error between v_{ot} and the Principal Component Analysis (PCA) subspace created by earlier observations, defined by Eq. 4.16 as follows,

$$y_e = exp\left(-\left\|(v_{ot} - \bar{v}_o) - UU^T\,(v_{ot} - \bar{v}_o)\right\|^2\right) \tag{4.16}$$

Note that the distance y_e is a Gaussian distribution, with Eigen vectors U and mean $\bar{v}_o$, $y_e = p\,(v_{ot}|p) \propto J\left(v_{ot}; \bar{v}_o, UU^T + \varepsilon I\right)$ as $\varepsilon \to 0$, and the inverse matrix can be solved by applying the Woodbury formula [16], given in Eq. 4.17 as follows,

$$(UU + \varepsilon I)^{-1} = \varepsilon^{-1}\left(I - (1+\varepsilon)^{-1}\,UU^T\right) \tag{4.17}$$

The optimal parameter p^{*} is found with a number of iterations. Here, empirical gradient is used, since the cost function is evaluated in the neighbourhood of the current parameter vector value. The tracking algorithm used here works as follows,

1. Manually locate target object in the first frame (t = 1). Eigen vectors U are initialized as empty. The tracker initially works as a template based tracker.
2. At the next frame, find the optimal parameters $p^{*} = argmax\left(\left\{y_e\left(p_i^{\omega} *\right)\right\}\right)$ over a number of iterations:

   - For each parameter p_i.
   - For each Δp and $\omega \in \{-1, 1\}$, compute $p_i^{\omega}\,(p_1, \ldots, p_i + c\Delta p, \ldots, p_{c+4})$
   - Compute $i^{*} = max\left\{y_e\left(p_i^{\omega}\right)\right\}$
   - Do $p \leftarrow p_i *$, store $y_e\left(p_i^{\omega} *\right)$

Fig. 4.4 Demonstration of tracker running on an image sequence from the GRID audiovisual corpus. Selected frames from one sequence are shown. This figure demonstrates the automatic movement of the ROI to maintain focus on the lip region

3. Check the observation vector: $v_o = v_o\left(\Xi\left(S_e^{'}, p^*\right)^{-1}\right)$ where Ξ is a transformation matrix, with result estimation phase as shown in Eq. 4.12.
4. If $y(v_o) = 1$, this signifies a good result to add to the model. When the desired number of new images has been accumulated, perform an incremental update.
5. Return to step 2.

The performance of this tracker is shown in Fig. 4.4, which shows selected frames from one image sequence. The rectangle around the lip region represents the ROI, which was identified in the first frame, and subsequent frames then automatically tracked this region to maintain focus on the desired area.

After tracking a sequence of lip images with this technique, the 2D-DCT vector $F_v = 2D - DCT(v)$ of each image in the sequence is found. A number of different visual feature extraction techniques have been used in the literature, but as shown in Chap. 2, 2D-DCT is extremely common and has been used by others, such as [18, 19]. DCT was originally developed in 1974 by [20], and is a close relative of the DFT. This was extended for application with image compression by [21]. The one-dimensional DCT is capable of processing one-dimensional signals such as speech waveforms. However, for analysis of two dimensional signals such as images, a 2D-DCT version is required. For a $V_U \quad x \quad V_V$ matrix V_P of pixel intensities, the

2D-DCT is computed in a simple way: the 1D-DCT is applied to each row of V_P and then to each column of the result. Thus, the transform of V_P is given by the DCT matrix V^{DCT},

$$V_{m,n}^{DCT} = V_{Wm} V_{Wn} \sum_{Vu=0}^{V_U-1} \sum_{Vv=0}^{V_V-1} V_{Pvu,vv} \cos\left(m(2V_u+1)\cdot\frac{\pi}{2V_u}\right) \cos\left(n(2V_v+1)\cdot\frac{\pi}{2V_v}\right) \tag{4.18}$$

with $0 \le m \le V_U - 1, 0 \le n \le V_V - 1$,

$$V_{Wn} = \begin{cases} \sqrt{1/V_v} & \text{if } n = 0 \\ \sqrt{2/V_v} & \text{otherwise} \end{cases} \text{ and } V_{Wm} = \begin{cases} \sqrt{1/V_U} & \text{if } m = 0 \\ \sqrt{2/V_U} & \text{otherwise} \end{cases} \tag{4.19}$$

Since the 2D-DCT can be computed by applying 1D transforms separately to the rows and columns, this means that the 2D-DCT is separable in the two dimensions. The first 30 2D-DCT components of each image are vectorised in a zigzag order to produce the vector for a single frame in an image sequence. The resulting 2D-DCT sequence of frames is then interpolated to match the equivalent audio log filterbank matrices for the matching speech sentence. As the video used in this work is recorded at 25 fps, this means that the 2D-DCT sequence is upsampled to match the audio features by using the same visual feature frame for four consecutive audio frames, a technique commonly used in the literature.

4.6 Visually Derived Wiener Filtering

Wiener filtering [22] is a signal processing technique that aims to clean up a noisy signal by comparing a noisy input signal with an estimation of a noiseless signal. One challenging aspect of Wiener filtering is the acquisition of an estimation of the noiseless signal. Unlike some other speech filtering approaches, some knowledge of the original signal is required. Visual information is used as the means of producing the estimate of the original audio signal, and this estimate is compared to the noisy signal. This represents the first stage of filtering in this two-stage approach and acts as the pre-processing step before the audio-only beamforming described later in this chapter. The Fourier transformed audio signal is used as an input, in tandem with associated visual information. In this system, the Wiener filter, $W(\gamma)$, is calculated in the frequency domain from the Power Spectrum (PS) estimate of clean speech ($\Psi_{\hat{a}}(\gamma)$) and the noisy speech mixture PS ($\Psi_a(\gamma)$) as,

$$W(\gamma) = \frac{\Psi_{\hat{a}}(\gamma)}{\Psi_a(\gamma)} \tag{4.20}$$

This produces the Wiener filter to be applied to the input signal. However, as stated, this is challenging to implement in this form, due to the difficulty of accurately calculating the clean PS, $\Psi_{\hat{a}}(\gamma)$, from visual information. Firstly, there are a variety of ways to calculate the PS of the noisy signal. One example is utilised by [23], where a VAD is used to identify non speech frames, and a noise alone PS is calculated. This is then added to the estimated speech alone PS to produce an estimated noisy speech PS. However, in this work, it was decided to use the power spectrum of the noisy speech mixture as a whole. This is a parameter that can be varied without difficulty, and using the noisy PS on a frame by frame basis also allows for a wide frame by frame variation in potential volume or type (aircraft, white noise etc.) of noise source. So therefore, $\Psi_a(\gamma)$ is simple to calculate from the noisy audio signal, but it is less straight forward to estimate the noise free PS.

This is where the input data from visual feature information can be utilised. Although it is very hard to estimate PS information directly from visual information, it is possible to estimate log filterbank values. Therefore, in the system, it is proposed to make use of the log filterbank vectors $F_a(1), \ldots, F_a(4)$, as described in Sect. 4.4, and the 2D-DCT vector F_v, which is calculated as outlined in Sect. 4.5, as inputs into the filter, with each audio channel being processed separately. Previous work by others [5, 24] showed that it is possible to estimate audio features from visual features, and produce the estimated noise free log filterbank vectors $F_{\hat{a}}(1), \ldots, F_{\hat{a}}(4)$. The production of these estimates is described in more depth in Sect. 4.7. For each channel, this is then transformed into a linear filterbank estimate of the clean audio signal, which is then interpolated to match the dimensionality of the audio 2D-DCT $\Psi_a(\gamma)$ with pchip interpolation [25]. This produces an estimate of the noise free power spectrum, $\Psi_{\hat{a}}(\gamma)$, which can be used to find $W(\gamma)$ as shown in Eq. 4.20. To find the enhanced power spectrum value, $\Psi_{\bar{a}}(\gamma)$, the noisy power spectrum $\Psi_a(\gamma)$ and the Wiener filter $W(\gamma)$ can be used as given in Eq. 4.21,

$$\Psi_{\bar{a}}(\gamma) = \Psi_a(\gamma)W(\gamma) \tag{4.21}$$

The key aspect of Eq. 4.20 is producing an estimate of the clean audio filterbank signal $F_{\hat{a}}$ to use as part of the filter. In this work, it is proposed to make use of Gaussian Mixture Regression (GMR), as described by [26], and outlined in Sect. 4.7. Following this filtering, the phase, $\varpi(\gamma)$, of each F_a is calculated and combined with $\Psi_{\bar{a}}(\gamma)$, to update the frequency domain Fourier transform $\mathbf{Z}(k, l)$ (see Eq. 4.3), for further processing.

4.7 Gaussian Mixture Model for Audiovisual Clean Speech Estimation

As mentioned in the previous section, a crucial aspect of this system is the production of an estimate of the noise free signal for use by the Wiener filter. In order to provide such an estimate, the joint audio and visual speech relationship has to be modelled. There are a variety of different approaches, for example, it is possible to make use of an approach utilising GMMs as demonstrated by [23]. There are also a range of other modelling alternatives available, such as using a number of different GMMs for each speech phoneme, requiring significant speech segmentation both in training and in the actual system. One alternative to this approach though, is one that was developed by [26], Gaussian Mixture Model-Gaussian Mixture Regression (GMM-GMR). This is a technique that was originally developed for robot arm training, and in this work, it has been adapted to be applied to the estimation of log filterbank audio vectors given a training set for offline training and valid visual input data. Although the Maximum a Priori (MAP) approach has been used in this field previously, to the knowledge of the author, this research represents the first example of applying this GMM-GMR technique to audiovisual data for speech filtering. The performance of this approach is discussed in Chap. 5.

To implement this GMM-GMR approach, this work makes use a method first outlined by [26] to encode the audiovisual signals in a mixture of GMMs, by considering each visual DCT vector F_v as an input in order to find an estimation of the equivalent noiseless audio signal by using GMR.

A mixture model of Q components of the joint audiovisual vector F_{av} is defined by a Probability Density Function (PDF),

$$e(F_{av}) = \sum_{q=1}^{Q} e(q)e(F_{av}|q) \tag{4.22}$$

with $e(q)$ representing the prior, and $e(F_{av} \mid q)$ representing the conditional PDF.

To model the joint audiovisual data F_{av} of dimension C, an offline training set is needed to train a mixture of Q Gaussians of dimensionality C. The performance of this aspect of the system is dependent on the training data provided, and so a training set using the GRID Corpus was used for this purpose, combining audio and visual data into a single training set. The detailed composition of the training set is discussed in Chap. 5. Returning to the PDF described in (4.22), the parameters in Eq. 4.22 become,

$$\begin{aligned} e(q) &= \pi_q \\ e(F_{av}|q) &= N(F_{av}; \mu_q, \Sigma_q) \\ &= \frac{1}{\sqrt{(2\pi)^C |\Sigma_q|}} e^{\frac{1}{2}((F_{av}-\mu_q)^T \Sigma_q^{-1} (F_{av}-\mu_q))} \end{aligned} \tag{4.23}$$

with π_q representing the prior, μ_q the mean, and Σ_q the covariance matrix of Gaussian component q. K-means clustering is applied to the joint vector training set to produce an initial estimate of GMM parameters, and Maximum Likelihood Estimation is performed on the model using Expectation Maximisation. The trained GMMs can then be used to perform GMR and with the aid of the input visual vector F_v, return an estimated value of the noiseless filterbank audio vector. For each frame of the speech signal, the mean and the covariance matrix of the Gaussian component q are divided into their visual and audio components, as defined by,

$$\mu_q = \{\mu_{v,q}, \mu_{a,q}\} \quad , \quad \Sigma_q = \begin{pmatrix} \Sigma_{v,q} & \Sigma_{va,q} \\ \Sigma_{av,q} & \Sigma_{a,q} \end{pmatrix} \tag{4.24}$$

For each Gaussian component q, $\hat{\Sigma}_{a,q}$, the expected conditional covariance, of $F_{a,q}$ given F_v is defined as,

$$\hat{\Sigma}_{a,q} = \Sigma_{a,q} - \Sigma_{av,q}(\Sigma_{v,q})^{-1}\Sigma_{va,q} \tag{4.25}$$

and $F_{\hat{a},q}$, the conditional expectation of $F_{a,q}$ given F_v is defined as,

$$F_{\hat{a},q} = \mu_{a,q} + \Sigma_{av,q}(\Sigma_{v,q})^{-1}(F_v - \mu_{v,q}) \tag{4.26}$$

$F_{\hat{a},q}$ and $\hat{\Sigma}_{a,q}$ are mixed depending on the probability that $q \in \{1, \ldots, Q\}$ has of being responsible for F_v, as shown by,

$$\beta_q = \frac{e(F_v|q)}{\Sigma_{i=1}^{Q} e(F_v|q)} \tag{4.27}$$

For a mixture of Q components, $\hat{\Sigma}_a$, the conditional covariance, and $F_{\hat{a}}$, the conditional expectation of F_a given F_v are defined as,

$$\hat{\Sigma}_a = \sum_{q=1}^{Q} \beta_q^2 \hat{\Sigma}_{a,q}, \quad F_{\hat{a}} = \sum_{q=1}^{Q} \beta_q F_{\hat{a},q} \tag{4.28}$$

Where $F_{\hat{a}}$ represents the estimated log filterbank signal to be processed further as described in Sect. 4.6. This audio log filterbank estimation is then used as part of the visually derived filtering approach, and enables the first stage of the two-stage speech filtering process to be performed. The resulting filtered signals are then used for audio-only beamforming.

4.8 Beamforming

Multiple microphone techniques such as beamforming can improve the quality and intelligibility of speech by exploiting the spatial diversity of speech and noise sources to filter speech. This is an active research field, with many different techniques developed. Within these techniques, one can differentiate between fixed and adaptive beamformers. The former combines the noisy signals by a time-invariant filter-and-sum operation, the latter combine the spatial focusing of fixed beamformers with adaptive noise suppression, such that they are able adapt to changing acoustic environments and generally exhibit a better noise reduction performance than fixed beamformers. The Generalised Sidelobe Canceller (GSC) is a very widely used structure for adaptive beamformers and a number of algorithms have been developed based on it. Among them, the general Transfer Function Generalised Sidelobe Canceller (TFGSC) suggested by [27], has shown impressive noise reduction abilities in a directional noise field, while maintaining low speech distortion.

In this work, the TFGSC beamformer is used on the pre-processed speech and noise mixtures as extracted in Sect. 4.3 and pre-processed in Sect. 4.6. This single modality technique receives multiple microphone signals, and then utilises them to output a single filtered signal. This follows examples of directional microphones utilised in commercial hearing aids and multi microphone array listening aids, as summarised in Chap. 3. In the system presented in this chapter, the input signals have been pre-processed by the visually derived Wiener filtering before being processed by audio-only beamforming. The beamforming approach used here is loosely integrated into the system, and so can be replaced by a different filtering mechanism to take account of further state-of-the-art research developments.

The general GSC structure is composed of three main parts: a Fixed Beamformer (FBF) $\mathbf{G}(k)$, a Blocking Matrix (BM) $\bar{\mathbf{G}}(k)$, and a multichannel Adaptive Noise Canceller (ANC) $\mathbf{H}(k, l)$. The FBF is an array of weighting filters that suppresses signals arriving from unwanted directions. The column of the BM can be regarded as a set of spatial filters suppressing any component impinging from the direction of the signal of interest, thus yielding $M - 1$ reference noise signals $\mathbf{\Lambda}(k, l)$. These signals are used by the ANC to construct a noise signal to be subtracted from the FBF output. This technique attempts to eliminate stationary noise that passes through the fixed beamformer, yielding an enhanced output signal $\bar{X}(k, l)$. Thus, the enhanced beamformer output $\bar{X}(k, l)$ can be written as,

$$\bar{X}(k, l) = \bar{X}_{FBF}(k, l) - \bar{X}_{NC}(k, l) \tag{4.29}$$

where $\bar{X}_{FBF}(k, l)$ represents the output of the *FBF*, and $\bar{X}_{NC}(k, l)$ the noise signal to be subtracted from the *FBF*. The *FBF* output can be described as,

$$\bar{X}_{FBF}(k, l) = \mathbf{G}^H(k, l)\mathbf{Z}(k, l) \tag{4.30}$$

with $\mathbf{G}^H(k, l)$, representing the *FBF*, and $\mathbf{Z}(k, l)$, the Fourier transformed microphone inputs, as described in Sect. 4.3. The noise signal to be subtracted from this, $\bar{X}_{NC}(k, l)$, is defined as thus,

$$\bar{X}_{NC}(k, l) = \mathbf{H}^H(k, l)\mathbf{\Lambda}(k, l) \tag{4.31}$$

where $\mathbf{H}^H(k, l)$, represents the ANC, value and $\mathbf{\Lambda}(k, l)$ the reference noise signals, defined as,

$$\mathbf{\Lambda}(k, l) = \bar{\mathbf{G}}^H(k, l)\mathbf{Z}(k, l) \tag{4.32}$$

with $\bar{\mathbf{G}}(k)$ being the BM and $\mathbf{Z}(k, l)$, as already mentioned, being the Fourier transformed microphone inputs. The *FBF* and BM matrices are constructed using the acoustical transfer function (ATF) ratios as follows,

$$\mathbf{G}(k, l) = \frac{\mathbf{A}(k, l)}{\|\mathbf{A}(k, l)\|^2} \tag{4.33}$$

$$\bar{\mathbf{G}}(k, l) = \begin{bmatrix} -\frac{A_2^*(k,l)}{A_1^*(k,l)} & -\frac{A_3^*(k,l)}{A_1^*(k,l)} & \cdots & -\frac{A_M^*(k,l)}{A_1^*(k,l)} \\ 1 & 0 & \cdots & 0 \\ 0 & 1 & \ddots & 0 \\ 0 & 0 & \cdots & 1 \end{bmatrix} \tag{4.34}$$

Note that the computation of both $\mathbf{G}(k)$ and $\bar{\mathbf{G}}(k)$ requires the knowledge of the ATF ratios. In this work, for simplicity, the true impulse responses $a_m(t)$, are directly transformed as defined in Sect. 4.3, into the frequency domain. This filtering operation produces the output frequency domain filtered output signal $\bar{X}(k, l)$. To transform this signal back to the time domain, an inverse Fourier transform is carried out, resulting in $\bar{x}(t)$, the production of the final output signal of the two-stage filtered speech system. This final output is then used for performance analysis of the system, as described in Chap. 5.

References

1. R. Zelinski, A microphone array with adaptive post-filtering for noise reduction in reverberant rooms, in *1988 International Conference on Acoustics, Speech, and Signal Processing, ICASSP-88*, (1988), pp. 2578–2581
2. T. Van den Bogaert, S. Doclo, J. Wouters, M. Moonen, Speech enhancement with multichannel Wiener filter techniques in multimicrophone binaural hearing aids. J. Acoust. Soc. Am. **125**, 360–371 (2009)
3. B. Rivet, L. Girin, C. Jutten, Log-Rayleigh distribution: a simple and efficient statistical representation of log-spectral coefficients. IEEE Trans. Audio Speech Lang. Process. **15**(3), 796–802 (2007)
4. A. Hussain, S. Cifani, S. Squartini, F. Piazza, T. Durrani, A novel psychoacoustically motivated multichannel speech enhancement system, *Verbal and Nonverbal Communication Behaviours*, (2007), pp. 190–199
5. I. Almajai, B. Milner, J. Darch, S. Vaseghi, Visually-derived Wiener filters for speech enhancement, in *IEEE International Conference on Acoustics, Speech and Signal Processing, ICASSP 2007*, vol. 4, (2007), pp. 585–588
6. Q. Nguyen, M. Milgram, Semi adaptive appearance models for lip tracking, in *ICIP09*, (2009), pp. 2437–2440
7. A. Abel, A. Hussain, Q. Nguyen, F. Ringeval, M. Chetouani, M. Milgram, Maximising audio-visual correlation with automatic lip tracking and vowel based segmentation, in *Proceedings of Biometric ID Management and Multimodal Communication: Joint COST 2101 and 2102 International Conference, BioID_MultiComm 2009, Madrid, Spain, 16–18 September, 2009*, vol. 5707, (Springer, 2009), pp. 65–72

8. Y. Cheung, X. Liu, X. You, A local region based approach to lip tracking. Pattern Recognit. **45**, 3336–3347 (2012)
9. T. Cootes, G. Edwards, C. Taylor, Active appearance models, in *Computer Vision-ECCV'98* (1998), pp. 484–498
10. G. Iyengar, G. Potamianos, C. Neti, T. Faruquie, A. Verma, Robust detection of visual ROI for automatic speechreading, in *2001 IEEE Fourth Workshop on Multimedia Signal Processing*, (IEEE, 2001), pp. 79–84
11. T. Wark, S. Sridharan, A syntactic approach to automatic lip feature extraction for speaker identification, in *Proceedings of the 1998 IEEE International Conference on Acoustics, Speech and Signal Processing*, vol. 6, (IEEE, 1998), pp. 3693–3696
12. P. Viola, M. Jones, Rapid object detection using a boosted cascade of simple features, in IEEE computer society conference on computer vision and pattern recognition, IEEE Comput. Soc. **1**, 511–518 (2001)
13. D.-J. Kroon, Viola jones object detection (2010). http://www.mathworks.com/matlabcentral/fileexchange/29437-viola-jones-object-detection
14. G. Bradski, The OpenCV Library. Dr. Dobb's J. Softw. Tools **25**, 120–126 (2000)
15. A. Levey, M. Lindenbaum, Sequential Karhunen-Loeve basis extraction and its application to images. IEEE Trans. Image Process. **9**(8), 1371–1374 (2000)
16. G. Golub, C. Van Loan, *Matrix Computations* (Johns Hopkins University Press, Baltimore, 1996)
17. G. Cauwenberghs, T. Poggio, Incremental and decremental support vector machine learning, in *Proceedings of the 2000 Conference, Advances in Neural Information Processing Systems*, vol. 13, (The MIT Press, 2001), pp. 409–415
18. M. Sargın, E. Erzin, Y. Yemez, A. Tekalp, Lip feature extraction based on audio-visual correlation, in *Proceedings of EUSIPCO*, vol. 2005 (2005)
19. I. Almajai, B. Milner, Maximising audio-visual speech correlation, in *Proceedings of AVSP* (2007)
20. N. Ahmed, T. Natarajan, K. Rao, Discrete cosine transform. IEEE Trans. Comput. **100**(1), 90–93 (1974)
21. W. Chen, W. Pratt, Scene adaptive coder. IEEE Trans. Commun. **32**(3), 225–232 (1984)
22. N. Wiener, *Extrapolation, Interpolation, and Smoothing of Stationary Time Series: With Engineering Applications* (The MIT Press, Cambridge, 1949)
23. I. Almajai, B. Milner, Effective visually-derived Wiener filtering for audio-visual speech processing, in *Proceedings of Interspeech, Brighton, UK* (2009)
24. M. Sargin, Y. Yemez, E. Erzin, A. Tekalp, Audiovisual synchronization and fusion using canonical correlation analysis. IEEE Trans. Multimed. **9**(7), 1396–1403 (2007)
25. F. Fritsch, R. Carlson, Monotone piecewise cubic interpolation. SIAM J. Numer. Anal. **17**(2), 238–246 (1980)
26. S. Calinon, F. Guenter, A. Billard, On learning, representing, and generalizing a task in a humanoid robot. IEEE Trans. Syst. Man Cybern. Part B **37**(2), 286–298 (2007)
27. S. Gannot, D. Burshtein, E. Weinstein, Signal enhancement using beamforming and nonstationarity with applications to speech. IEEE Trans. Signal Process. **49**(8), 1614–1626 (2001)

Chapter 5
Experiments, Results, and Analysis

Abstract As discussed in Chap. 2, the multimodal nature of both human speech production and perception is well established. The work presented in this book has utilised this multimodal speech relationship to design a two-stage multimodal speech filtering system, making use of both audio and visual information. This chapter presents a detailed investigation of system performance in several different audio-visual scenarios. The first experiment evaluates the speech filtering performance of the system in a range of very noisy environments. This represents an example of a challenging real world situation that a potential user of a speech enhancement system may encounter, with speech from the GRID corpus mixed with aircraft cockpit noise at a variety of different SNR levels, ranging from –40 dB to +10 dB. A number of potential strengths and limitations are also discussed. Overall, this chapter will evaluate the performance of the two-stage multimodal system presented in this work, using both objective listening tests and subjective evaluation by human listeners to identify the strengths and limitations of the system. The discussion will also identify areas in which these limitations can be overcome.

Keywords Audiovisual · Multimodal · Experiments · Wiener filtering · Speech enhancement

5.1 Speech Enhancement Evaluation Approaches

To evaluate the performance of the two-stage multimodal speech enhancement system presented in this work, a variety of measures are used. These come in two forms, objective and subjective. Objective measures are those that are calculated automatically by machine. These have the benefit of being faster to perform than subjective listening tests. There are many different approaches with regard to objective tests such as the Perceptual Evaluation of Speech Quality (PESQ) [1] measure, the Itakura-Saito Distance (IS) [2, 3], or SNR level gain. Subjective testing is also used for the evaluation performed in this chapter, which is carried out with the use of human volunteers, and is widely regarded to be the most accurate way of evaluating the performance of speech enhancement algorithms [4]. In these tests, listeners hear filtered speech sentences and score each sentence.

A. Abel and A. Hussain, *Cognitively Inspired Audiovisual Speech Filtering*,
SpringerBriefs in Cognitive Computation, DOI 10.1007/978-3-319-13509-0_5

5.1.1 Subjective Speech Quality Evaluation Measures

Subjective listening tests use volunteers to evaluate speech. Many different approaches have been used, such as word identification in a sentence [5], and scoring the overall quality of speech according to the opinion of the listener. However, the number of different approaches makes comparison of results between different publications more difficult, and a simple measure of overall speech quality does not always provide a comprehensive picture of listener opinion. Many speech enhancement algorithms introduce distortion as well as removing noise, and so this should be taken into account in listening tests. In recent years, a standardised approach has been developed by the International Telecoms Union (ITU-T), (ITU-T recommendation P. 835 [6]), and utilised in reviews of objective test measures [4] and recent work by [7]. This approach requires the listener to listen to each sentence, and then score it from one to five based on three criteria. Firstly, a score for speech distortion level is recorded, with 1 indicating the most, and 5 indicating the least distortion. Secondly, the listener gives a score (again between 1 and 5, with 5 indicating the least noise intrusiveness) for the level of noise intrusiveness, before finally giving a score for the overall speech quality. These three scores are used to produce MOS for each evaluation measure.

In the listening tests reported in the remainder of this chapter, nine volunteers participated, and each volunteer heard sentences from the test-set at six different SNR levels (–40 dB to +10 dB). Six of the nine volunteers were male and three were female, all with a good level of hearing and all spoke English fluently (six of the volunteers spoke English as a first language, three did not). All volunteers were postgraduate research students, although none were speech processing specialists. These tests took place in a soundproofed room using headphones, and each sentence was played randomly to listeners, who then assigned a score from 1 to 5, with 1 being worst and 5 best, using the three criteria discussed above (speech signal distortion, noise intrusiveness level, and overall speech quality). For purposes of comparison, three versions of each sentence were played. The noisy sentence with no speech processing, the sentence processed with an audio-only spectral subtraction approach [8], and thirdly, the sentence processed with the audiovisual approach proposed here.

5.1.2 Objective Speech Quality Evaluation Measures

There are a number of issues with the use of subjective testing. Comprehensive listening tests can be time consuming and dull, leading to listener fatigue, and it can be difficult to find an adequate number of suitable volunteers [9]. It can often be useful to run objective tests in addition to listening tests. Objective tests are carried out automatically by machine rather than by using the subjective opinion of human volunteers. These have the advantage of being much quicker to conduct, and to varying extents, can correlate and confirm the results of subjective tests. Many different measures have been devised to objectively assess the performance of speech enhancement algorithms. Work by [2, 4, 9] has focused on the evaluation of many of these measures, some of which were not originally designed for assessing the performance of speech enhancement. The development and testing of composite measures is also covered in work by [2], combining a number of objective measures into a single, theoretically more accurate measure.

One objective measure that is very widely used is the PESQ [1] algorithm. This has been recommended by ITU-T recommendation P. 862 [10] for measurement of narrow band telephony related speech enhancement. It compares a clean reference signal to the speech signal to be evaluated and returns a score ranging from –0.5 to 4.5, which means it can be compared to subjective MOS results. This algorithm is described in more detail in [11]. Both clean and noisy signals are aligned and transformed in a manner that simulates human hearing, and disturbance processing is carried out to look for errors in the signal being tested.

Another objective measure that is used in this work is Weighted-Slope Spectral Distance (WSS) [4, 12], another full reference measure. WSS functions by comparing the spectral slope distance in each spectral band. The spectral slope refers to the distance between adjacent spectral magnitudes, and is measured in decibels. Segmental SNR (SegSNR) is a time domain objective measure, and functions by averaging SNR level estimates from frame to frame [4]). The final individual objective measure used is the Log-Likelihood Ratio (LLR). This is a form of Linear Predictive Coding (LPC) based evaluation. LLR is defined in more detail in [4].

It is also possible to combine these measures to create unified composite speech evaluation techniques. [2, 4, 9] investigated the correlation between subjective listening tests and a wide range of objective measures, and found that the level of correlation between subjective and objective scores for different measures (speech/noise distortion, overall quality) varies depending on the subjective measure considered. They state that an individual objective measure is unlikely to correlate highly in all three aspects, and one conclusion from their work was that basic objective measures correlated very poorly with noise distortion. Three composite measures were then defined that combined the strongest individual objective measures described above, as the composite signal quality, background distortion, and overall score (C_{SIG}, C_{BAK}, and C_{OVL} respectively). These composite measures are defined by [2] as,

$$C_{SIG} = 3.093 - 1.029 \cdot LLR + 0.603 \cdot PESQ - 0.009 \cdot WSS$$

$$C_{BAK} = 1.634 + 0.478 \cdot PESQ - 0.007 \cdot WSS + 0.063 \cdot segSNR$$

$$C_{OVL} = 1.594 + 0.805 \cdot PESQ - 0.512 \cdot LLR - 0.007 \cdot WSS$$

5.2 Preliminary Experimentation

The main aspect of this system to be configured was the GMM-GMR audiovisual model component. This is used to calculate a smoothed estimate of the noise-free audio signal, based on matching visual information. There were two main parameters that required configuration. Firstly, the number of GMM components, and secondly, the ideal composition of the training set. Different combinations of training data were used for testing. These range from using 50 sentences from each speaker and creating a large 200 sentence dataset, to simply using 100 sentences from a single speaker. Different combinations were experimented with, with the hypothesis that using more than one speaker as part of the training set produces a more flexible model. The test-set used for this informal evaluation was relatively small and contained three sentences from each of the four speakers selected for use in this work, creating a 12 sentence test-set. In addition to these two variables, a further factor to consider is the performance at different SNR levels. Therefore, the combination of different possible system configurations was also tested at three different SNR levels, –50 dB, –20 dB, and 0 dB, to simulate a range of noisy and less noisy environments.

The PESQ approach described in Sect. 5.1 is used. This approach produces a simple output and is quick to calculate. It was felt that at this preliminary stage, simple objective results would produce an adequate picture of overall performance. However, in order to obtain a more informed opinion, informal listening was also used.

A series of evaluations showed that for each SNR level considered, models trained with the use of 12 components produced the best results. With regard to the most suitable training sets, it was found that the most consistent and effective training sets consisted of either all four speakers, or combining two speakers. Therefore, it was decided for the remainder of this work to make use of the four speaker training set.

5.3 Automated Lip Detection Evaluation

5.3.1 Problem Description

To successfully exploit audiovisual information, it is important that the appropriate visual ROI (in this case lip-region information) is correctly identified and tracked. Manual frame by frame identification of the ROI is time consuming and represents an impractical approach. In the work presented here, a lip tracker has been utilised in order to extract the lip-region automatically in each frame of an image sequence. In this work, a Viola-Jones [13] detector has been implemented to automatically identify the initial lip-region. This was described in more depth in Chap. 4.

5.3.2 Experiment Setup

To assess performance of the lip detector, images from both of the main corpora used in this work (GRID and VidTIMIT) are used. Six speakers from the GRID corpus are used, along with four speakers from the VidTIMIT database. These speakers are split by gender (six male speakers, four female), as well as representing a number of different ethnicities, in order to test a range of speakers. This data was used to create several test-sets. The first test-set used images from both corpora, and consisted of the first image from a single sentence sequence, producing a small test-set of 10 images, to test the initial implementation of the detector and refine any problems that were identified. The second test-set created was used for the main evaluation of the detection approach. This consisted of sixty images from the ten speakers, chosen from the sequence at the start, and also at random (to provide different mouth shapes). A number of example images used in the test-set are shown in Fig. 5.1. The final test-set consists of 18 videos from 6 speakers from the GRID Corpus, used to inspect whether after identifying the correct ROI, the tracker could correctly use this location to track the correct location in subsequent frames. To evaluate the effectiveness of this approach (i.e. whether the final lip-region is identified correctly), the initial candidate locations identifed by the lip detector are visually compared to the final chosen ROI.

5.3.3 Results and Discussion

As described above, the results in this section were produced by visual inspection. The first dataset was used to identify bugs and test refinements. The lip detector was

Fig. 5.1 Example image frames from the GRID (*top* and *bottom left*) and VidTIMIT (*top* and *bottom right*) Corpora

initially found to successfully identify a range of possible ROI candidates and then select a suitable final ROI (based on the technique outlined in Chap. 4) in 9 of the 10 initial images. Examples of successfully detected ROI can be seen in Fig. 5.2. It can be seen that a number of possible mouth objects are detected (thinner rectangles), and then the final chosen ROI is correctly identified. One image was found to produce an incorrect result (shown in Fig. 5.3). Figure 5.3 shows that the problem with this image (confirmed by other informal tests on additional images to replicate the result) is the lack of detected candidates, rather than a problem with the final ROI. This is a problem with the standard trained model, and it was felt that to correct this would be outwith the scope of this research, due to the focus of this research not being exclusively on lip detection. As discussed in earlier chapters, training a Viola-Jones detector from can be very time consuming and intensive, and so the widely used standard Haar features are felt to be an acceptable compromise.

This problem was solved by checking the number of resulting potential candidates before outputting a final ROI. The solution was described in depth in Sect. 4.5, but to briefly summarise, the number of output candidate lip-regions was identified. If the number was below a threshold (manually defined in this work as 7), a full face detector is then run on the image (which was found to work without any problems in a variety of tests), and the mean face location is then cropped to only keep the

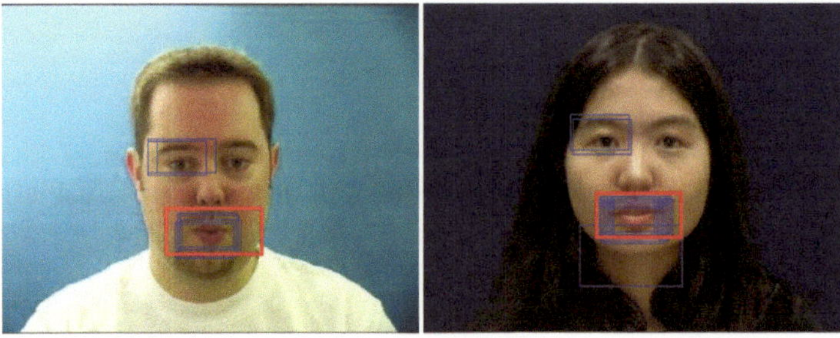

Fig. 5.2 Example of candidate lip-regions, indicated by *narrow rectangles*. The *thicker rectangle* shows the final chosen ROI

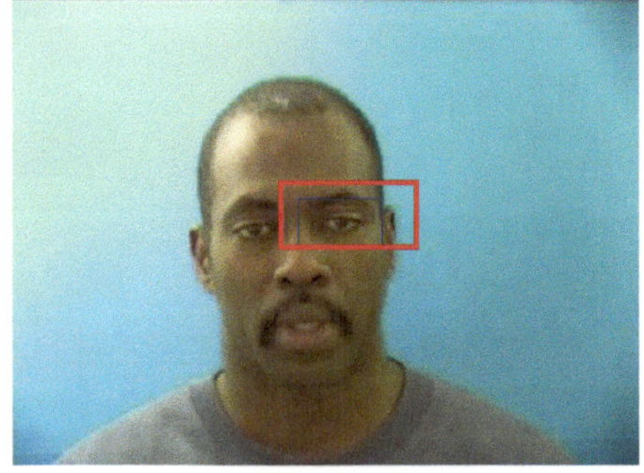

Fig. 5.3 Poor selection performance by the initial lip detection approach. It can be seen that only a single candidate location was identified (resulting in a narrow rectangle around the eye region), resulting in an incorrect final ROI output

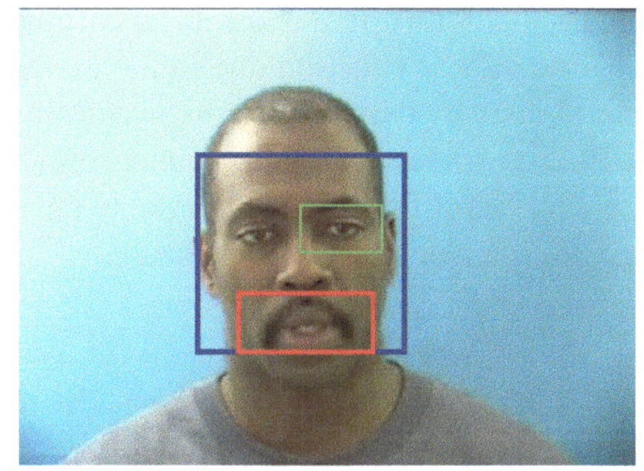

Fig. 5.4 Comparison of mean face detection location, compared to incorrect lip-region identification (*narrow rectangle*), showing final cropped lip-region output

approximate lip area. This was found to be a solution which resulted in error free results with the larger dataset.

After the refinement of the detector, the second set of 60 images was tested. All were visually inspected to check whether the final chosen ROI was appropriate, and also to observe the number of initial candidates. All images were found to identify the lip-region correctly. Where a sufficient number of candidates were identified, the correct ROI was chosen (as can be seen in Fig. 5.2), and in cases with a smaller number of candidates, then the face detector was run, and the automatic cropping identified an acceptable ROI, as shown in Fig. 5.4. The final tests involved checking if the detected image would allow for the image tracker (as described in Chap. 4) to function correctly. The video test-set described previously was tested by combining the lip detector and the tracker. In all of the 18 tested cases, the tracker functioned correctly with no problems and tracked the correct region.

However, there are some limitations with this approach. Firstly, the detector has only been tested with images using a single speaker in the frame. In order to identify the correct speaker if there is any conflicting visual information, further research would have to be performed, and an appropriate audiovisual VAD would have to be utilised. However, it is felt that this is outside the scope of this research. Also, this approach was tested only on data from the GRID and VidTIMIT corpora, meaning that while it functions well with these relatively clean corpora, it has not been fully tested in more difficult situations. Overall, it is felt that this represents an effective solution to the problem of automatically identifying the correct mouth ROI for lip tracking.

5.4 Noisy Audio Environments

5.4.1 Problem Description

One key challenge for speech enhancement algorithms is achieving performance in extremely noisy environments. A real world example of one such environment is on board an aircraft. In this environment, it can become very difficult for conventional hearing aids to function due to the extremely high level of background noise. The use of visually derived filtering in addition to conventional audio-only beamforming adds an extra level of speech enhancement capability and should theoretically allow for successful filtering in very noisy environments where conventional single stage speech filtering may perform badly.

5.4.2 Experiment Setup

To assess the performance of this system in extremely noisy environments, the multimodal approach described in Chap. 4 was tested with speech and noise mixtures that were combined in a simulated room environment. The parameters required by the audiovisual GMM were defined by preliminary experimentation. The number of components used in the GMM was set at 12, and the training set contained 200 sentences from four speakers from the GRID Corpus.

For testing, 61 different sentences were used. These sentences were different from those used in the training set, but made use of speakers that the audiovisual model had previously encountered. Each sentence was three seconds in length and when divided up into frames, produced 299 frames per sentence. The noise source was provided by using recorded F16 aircraft cockpit noise. These sources were mixed in the simulated room to produce the noisy speech mixture.

The composite measure (described in Sect. 5.1) is used. It was also felt that a more accurate approach than only using objective speech evaluation measures would be to also carry out subjective listening tests. Nine volunteers participated in these tests, and each volunteer was played sentences from the test-set at six different SNR levels (–40 dB to +10 dB). For purposes of comparison, the same three versions of each sentence as described above were played to the volunteers. This produced three MOS for the three different approaches, one for the overall score, one for speech distortion, and one for background noise intrusiveness. These scores were analysed and compared to the objective scores.

5.4.3 Results and Discussion

Composite measures are used in this work for objective evaluation of test sentences. The results for the Overall score (COvl), Background score (CBak), and Signal score (CSig) are shown in Figs. 5.5, 5.6, and 5.7, with data also displayed in Tables 5.1, 5.2, and 5.3. The enhanced audiovisual filtered speech (Avis) results were compared to noisy unfiltered speech (Noi), and speech filtered with audio only spectral subtraction (Spec).

Considering speech signal distortion first, it can be seen in Fig. 5.7 that at very low SNR levels, both unfiltered speech and spectral subtraction results produced

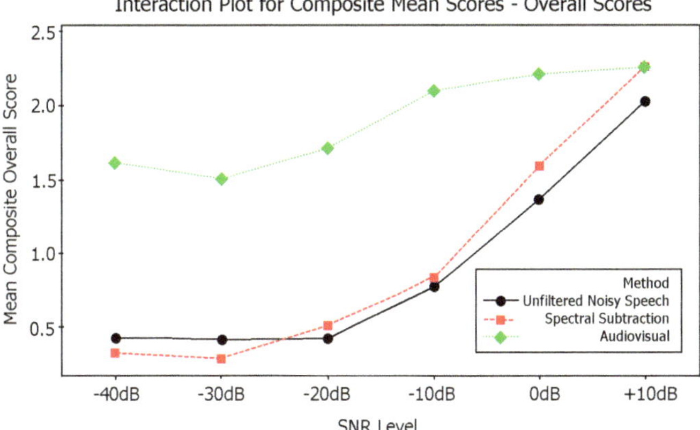

Fig. 5.5 Interaction plot for overall composite objective mean score at varying SNR levels, showing Unfiltered noisy speech, spectral subtraction, and audiovisual filtering scores

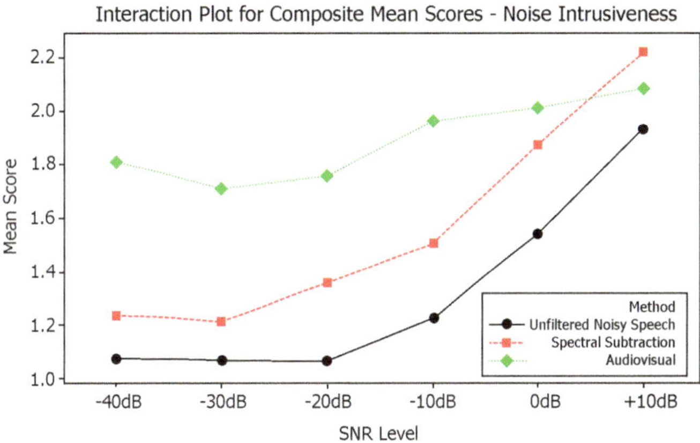

Fig. 5.6 Interaction plot for noise intrusiveness composite objective score at varying SNR levels, showing Unfiltered noisy speech, spectral subtraction, and audiovisual filtering scores

a small negative value at –40 dB and –30 dB, a very small result at –20 dB, and a very low positive result at –10 dB. This is because the testing algorithms were unable to identify an adequate level of speech in these results to assign a quality score. However, speech filtered with the audiovisual approach produced much better scores at these SNR levels, returning positive scores at all levels, increasing as the SNR level increased. At higher SNR levels (0 and +10 dB), spectral subtraction and unfiltered speech produced much improved results, with only a small, but statistically significant improvement ($p < 0.05$) over the unfiltered and audio only options seen when two-stage filtering is used. This can be seen more clearly in the interaction plot in Fig. 5.7, and the results of Bonferroni multiple comparison in Table 5.1. As it was

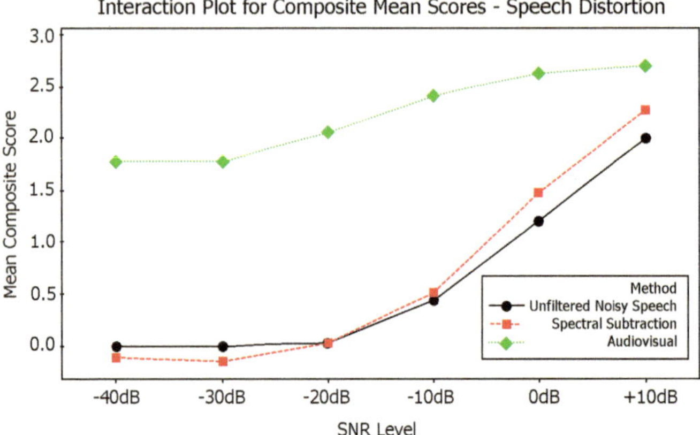

Fig. 5.7 Interaction plot for speech distortion composite objective score at varying SNR levels, showing Unfiltered noisy speech, spectral subtraction, and audiovisual filtering scores

Table 5.1 Selected results of Bonferroni multiple comparison, showing p-value results for difference between unfiltered speech and audiovisual filtering for speech distortion composite scores

Level (dB)	Difference of means	SE of difference	T-value	Adjusted P-value
–40	1.789	0.069	26.084	0
–30	1.786	0.069	26.035	0
–20	2.202	0.069	29.495	0
–10	1.971	0.069	28.737	0
0	1.415	0.069	20.627	0
+10	0.695	0.069	10.135	0

found that the results for unfiltered speech and audio only spectral subtraction were very similar, Table 5.1 focuses only on the p-values between unfiltered speech and audiovisual filtering.

The noise intrusiveness scores show slightly different results. The results show that there is significant improvement for noise intrusiveness at low SNR levels when using audiovisual filtering, as shown by the interaction plot in Fig. 5.6 and the selected p-values given in the results of Bonferroni multiple comparison in Table 5.3. The difference between the three scores is not as great as might be expected, and this difference tends to be lower than the signal distortion scores. At +10 dB, spectral subtraction slightly outperforms the audiovisual method. However, the most important scores to consider are the overall mean scores presented in Fig. 5.5. These show that at low SNR levels, the audiovisual approach significantly outperforms conventional spectral subtraction, as confirmed by selected Bonferroni multiple comparison results in Table 5.2. However, when there is less background noise present, the benefits are less obvious, with very similar overall scores for all three methods at +10 dB. This

Table 5.2 Selected results of Bonferroni multiple comparison, showing P-value results for difference between unfiltered speech and audiovisual filtering for overall composite scores

Level (dB)	Difference of means	SE of difference	T-value	Adjusted P-value
−40	1.184	0.066	17.974	0
−30	1.081	0.066	16.421	0
−20	1.282	0.066	19.475	0
−10	1.321	0.066	20.064	0
0	0.847	0.066	12.869	0
+10	0.230	0.066	3.497	0.075

Table 5.3 Selected results of Bonferroni multiple comparison, showing P-value results for difference between unfiltered speech and audiovisual filtering for noise intrusiveness composite scores

Level (dB)	Difference of means	SE of difference	T-value	Adjusted P-value
−40	0.734	0.037	19.660	0
−30	0.634	0.037	16.981	0
−20	0.690	0.037	18.457	0
−10	0.732	0.037	19.593	0
0	0.470	0.037	12.586	0
+10	0.153	0.037	4.107	0.007

suggests that the audiovisual filtering approach is most effective in extremely noisy environments, with relatively little improvement found in environments containing a lower level of noise.

To confirm the objective composite measure results above, subjective listening tests were used. As described in Sect. 5.1.1, nine volunteers participated in listening tests. Interaction plots of each MOS measure (overall score, speech quality, noise intrusiveness) are shown in Figs. 5.8, 5.9, and 5.10 respectively.

Firstly, looking at the overall results, it is clear that the listeners were consistently unable to hear unprocessed speech at very low SNR levels, and spectral subtraction was also of little use. However, in these noisy environments, the audiovisual approach produced higher scores, with the listeners able to identify speech. This pattern is mirrored for speech distortion levels, and also for noise, with the audiovisual approach demonstrating a large improvement at low SNR levels, showing that in very noisy environments, this two-stage approach can produce significantly improved results when it comes to speech quality and the overall score. This closely and accurately matches the results found with the composite measures. With regard to the significance of these results, it can be seen from the interaction plots that the mean scores for unfiltered speech and spectral subtraction are very similar, especially at very low SNR levels and so the focus is on the difference between unfiltered speech

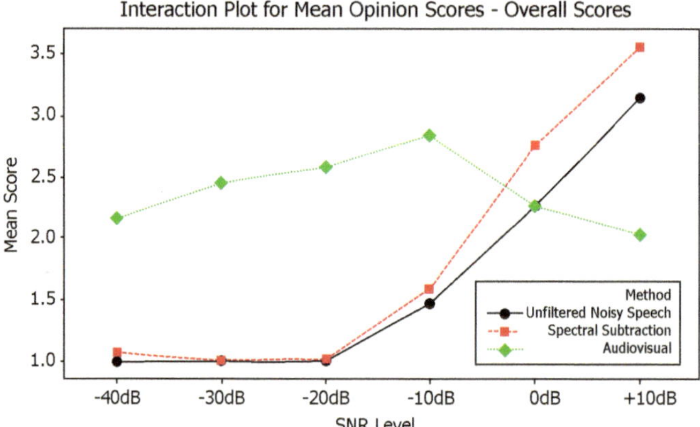

Fig. 5.8 Interaction plot for overall MOS score at varying SNR levels, showing unfiltered noisy speech, spectral subtraction, and audiovisual filtering scores

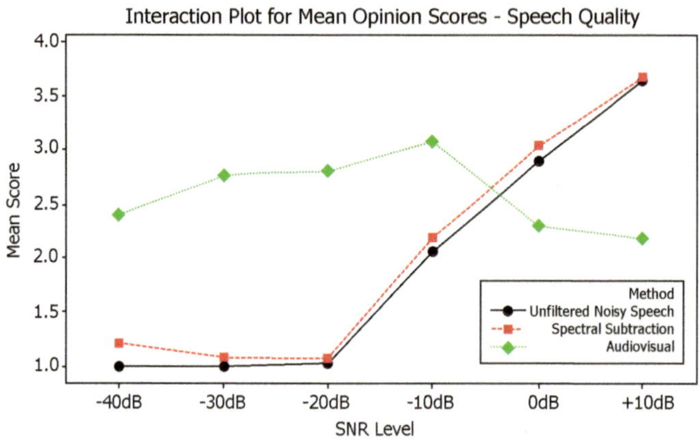

Fig. 5.9 Interaction plot for speech quality MOS score at varying SNR levels, showing unfiltered noisy speech, spectral subtraction, and audiovisual filtering scores

and audiovisual filtering. The relevant results of Bonferroni multiple comparison are summarised in Tables 5.4, 5.5 and 5.6, showing the difference between audiovisual filtering and unfiltered speech for each of the three MOS results, with p-values of $p < 0.05$ showing that the difference at low SNR levels is statistically significant for all three measures.

However, at higher SNR levels (0 dB and +10 dB), it can be seen that the overall score of the two-stage system is lower than at lower SNR levels, and this is especially noticeable in the signal quality level. Analysing the results in detail, it can be seen that spectral subtraction outperforms the multimodal approach at 0 dB, with listeners assigning a higher overall MOS to spectral subtraction, and a very similar score to noisy and two-stage filtered speech. At a SNR of +10 dB, the audiovisual method

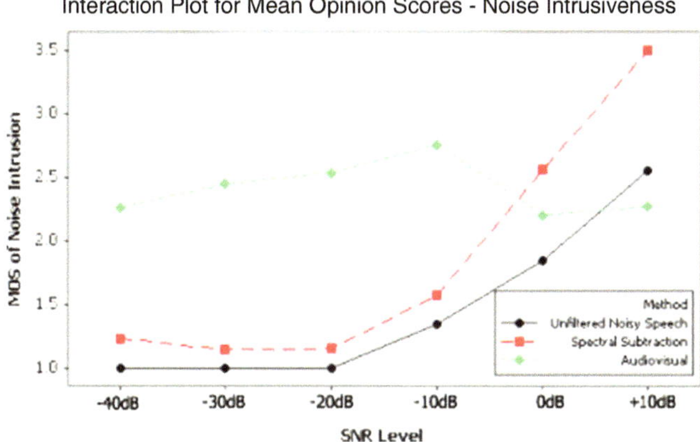

Fig. 5.10 Interaction plot for noise intrusiveness MOS score at varying SNR levels, showing unfiltered noisy speech, spectral subtraction, and audiovisual filtering scores

Table 5.4 Selected results of Bonferroni multiple comparison, showing P-value results for difference between unfiltered speech and audiovisual filtering for overall MOS scores

Level (dB)	Difference of means	SE of difference	T-value	Adjusted P-value
–40	1.167	0.736	15.837	0
–30	1.448	0.736	19.658	0
–20	1.585	0.736	21.519	0
–10	1.363	0.736	18.502	0
0	–0.015	0.736	–0.201	1
+10	–1.119	0.736	–15.180	0

Table 5.5 Selected results of Bonferroni multiple comparison, showing P-value results for difference between unfiltered speech and audiovisual filtering for speech quality MOS Scores

Level (dB)	Difference of means	SE of difference	T-value	Adjusted P-value
–40	1.396	0.732	19.084	0
–30	1.763	0.732	24.095	0
–20	1.778	0.732	24.298	0
–10	1.015	0.732	13.870	0
0	–0.600	0.732	–8.201	0
+10	–1.481	0.732	–20.250	0

performed very poorly, with a very low overall MOS in comparison to both noisy and spectral subtraction sentence scores. Looking at these results in detail, it can be seen that the main drop in audiovisual filtered MOS is displayed in the speech distortion

Table 5.6 Selected results of Bonferroni multiple comparison, showing P-value results for difference between unfiltered speech and audiovisual filtering for noise intrusiveness MOS scores

Level (dB)	Difference of means	SE of difference	T-value	Adjusted P-value
–40	1.259	0.083	15.152	0
–30	1.444	0.083	17.380	0
–20	1.533	0.083	18.450	0
–10	1.411	0.083	16.980	0
0	0.363	0.083	4.367	0.020
+10	–0.289	0.083	–3.476	0.079

score. The noise intrusiveness score for audiovisual filtering increases between –40 and –10 dB, but there is a small drop at 0 dB, followed by another drop at +10 dB. The most relevant result is the speech distortion score. It can be seen that at 0 dB, the audiovisual output speech quality is significantly reduced. The p-values in Table 5.4 show that there is no significant difference between the overall scores at 0 dB, but individual speech and noise scores are significantly different. At +10 dB, the speech quality score is extremely low in comparison to all other approaches, and Tables 5.4, 5.5 and 5.6 show that the audiovisual approach performs significantly worse. This suggests that the main reason for the low overall score is because of the level of speech distortion introduced, and listening to filtered sentences confirmed this. This represents a more accurate picture than presented by the objective results alone and shows that although objective measures are valuable, they need to be supplemented with listening tests.

One hypothesis is that this distortion is due to problems with the visual filterbank estimation approach. The approach evaluated in this paper, GMR, was originally designed to calculate efficient robot arm movement, and while it is shown in this work that it can be applied to audiovisual data, the resulting signal is not always accurately calculated, and so this technique can introduce distortion into the speech. Filtering at lower SNR levels produces good results, but when there is less background noise to remove, more distortion is introduced into the speech. A key solution to improving results in future work is to consider an improved audiovisual speech estimation model.

State-of-the-art work by [7] makes use of a MAP GMM approach and also further enhancements such as a VAD and using a number of different GMMs to represent individual speech phonemes. It was also shown in previous work by the author in [14] that a degree of asynchrony between audio and visual frames may also improve results, so there are many ways in which this initial audiovisual two-stage system can be improved in the future. While this two-stage approach produces improved results in very noisy environments, the use of visual information is not always helpful. The potential availability of audio and visual information in a more realistic speech environment also has to be considered. This work makes use of a simulated environment, with pre-recorded visual information and static speech and noise sources, but in a more realistic environment, such as one which a hearing aid might be expected to function in; there may be multiple inconsistent moving noise sources. Furthermore, visual information may not always be usable. Therefore, in addition to improving the visually derived filtering approach as described above, it is also important to consider how a multimodal system can best take advantage of audio and visual information to deliver good results on a frame by frame basis.

Overall, the results show that visual information can be used as part of a speech enhancement system. In very noisy environments, it can be seen that the two-stage speech enhancement system presented in this work is capable of successfully filtering speech, as proven by composite objective scores and subjective listening tests. While the poor scores at high SNR levels indicate that that the individual components of the system can still be refined further, for example, with a more sophisticated audiovisual speech model, the initial system has been shown to successfully filter speech in challenging environments.

5.5 Testing with Novel Corpus

5.5.1 Problem Description

An important limitation to consider is the performance of the system with speakers outside the range that it has been trained with. This is a common limitation, with considerable quantities of research [7, 15–17] explicitly being trained and tested with the same single speaker corpus.

5.5.2 Experiment Setup

The system was tested with the same parameters and configuration as defined in Sect. 5.4. To provide the speech source data, 66 sentences from the VidTIMIT audio-visual corpus were used for testing, which the system was completely untrained with. This corpus uses different speakers and content (TIMIT sentences). The noise source was provided by using F16 aircraft cockpit noise. Each test sentence was mixed with aircraft noise at SNR levels ranging from +10 dB to –40 dB. The composite measure described in Sect. 5.1 is used in this investigation to evaluate the filtered sentences. For purposes of comparison, three versions of each sentence were compared. The noisy sentence with no speech processing, the sentence processed with an audio only spectral subtraction approach [18], and the sentence processed with the audiovisual approach. The results were also compared to the results in Sect. 5.4.

Table 5.7 Selected results of Bonferroni multiple comparison of VidTIMIT corpus, showing P-value results for difference between unfiltered speech and audiovisual filtering for overall composite scores

Level (dB)	Difference of means	SE of difference	T-value	Adjusted P-value
–40	0.176	0.058	3.018	0.405
–30	–0.011	0.058	–0.190	1.000
–20	0.240	0.058	4.130	0.006
–10	1.123	0.058	19.277	0.000
0	1.958	0.058	33.619	0.000
+10	2.965	0.058	50.91	0.000

5.5.3 Results and Discussion

The composite results for the Signal score (CSig), Background score (CBak), and Overall score (COvl), were calculated. For reasons of space, as similar overall patterns for overall, speech, and noise results were found, only the overall results are discussed here, but full results have been published [19]. The overall difference of means is shown in Table 5.7. The enhanced audiovisual filtered speech (Avis) results were compared to noisy unfiltered speech (Noi), and speech filtered with audio only spectral subtraction (Spec). The Interaction plot is shown in Fig. 5.11.

It can be seen that the noisy unfiltered file produces the best overall results at all SNR levels. Table 5.7 shows that the difference is not significant at low SNR levels (–40 dB, –30 dB), but the unfiltered score is significantly better than audiovisual at higher SNR levels, which was confirmed by listening to the files. Table 5.7 shows that

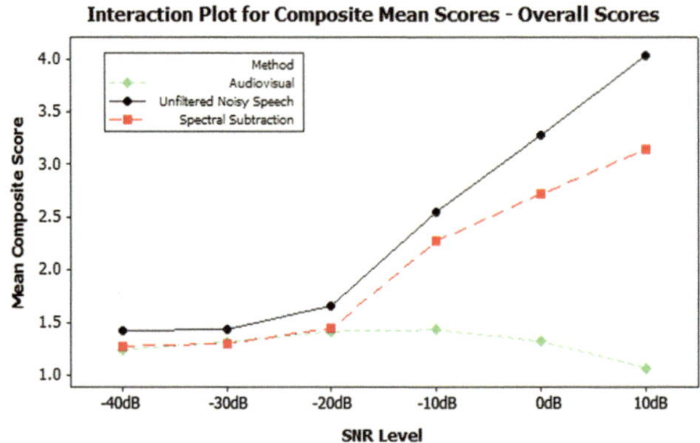

Fig. 5.11 Interaction plot for overall composite objective mean score of the VidTIMIT corpus at varying SNR levels, showing unfiltered noisy-speech, spectral subtraction, and audiovisual filtering scores

the unfiltered score also outperforms the spectral subtraction approach. The spectral subtraction approach was found to produce files that introduced some noticeable distortion, as confirmed by the difference between the scores, particularly at higher SNR levels. As stated, full findings are discussed in related work by the authors in [19].

There are several reasons for these findings. Firstly, the limitations with the audio-visual model, as discussed in Sect. 5.4 are repeated with the results presented in this section. Secondly, the VidTIMIT corpus is recorded in an environment with some background noise present. The objective measures work by performing a comparison of the processed file with the original file. If the original clean speech sentence contains noise, then this will affect the final scores, as shown here. Finally, as stated, these results also identify limitations that were not explored in similar work by others [7, 17].

In addition to the results presented above, a comparison of these results with the findings presented in Sect. 5.4 was also made. A comparison was also made of composite overall scores for sentences from the GRID and VidTIMIT corpora filtered using the audiovisual approach presented here. The results are shown in Fig. 5.12.

As was expected, taking into account the limitations of testing the audiovisual speech model on unseen data previously discussed in this section, and also the effect of the noisy corpus on overall scores, sentences from the GRID corpus were found to have a significantly higher mean score than sentences from the VidTIMIT Corpus. This was confirmed by the comparison of means in Table 5.8. Overall, these results confirm the limitation identified previously, that the system performs poorly when it is tested with data that is unrelated to the chosen training set. This is an expected result, as other similar research that uses a similar visually derived filtering technique such as work by [7] makes use of test data that closely matches the training set (i.e. the same single speaker corpus is used for training and testing). In order to improve this aspect of the system, an improved audiovisual model could be considered, making use of more sophisticated filtering and a much more varied training set.

Table 5.8 Selected results of Bonferroni multiple comparison of audiovisual filtered speech sentences, showing P-value results for difference between GRID corpus filtered speech and VidTIMIT corpus filtered speech for overall composite scores

Level (dB)	Difference of means	SE of difference	T-value	Adjusted P-value
–40	–0.372	0.086	–4.327	0.001
–30	–0.192	0.086	–2.231	1.000
–20	–0.299	0.086	–3.478	0.035
–10	–0.666	0.086	–7.750	0.000
0	–0.892	0.086	–10.390	0.000
+10	–1.186	0.086	–13.81	0.000

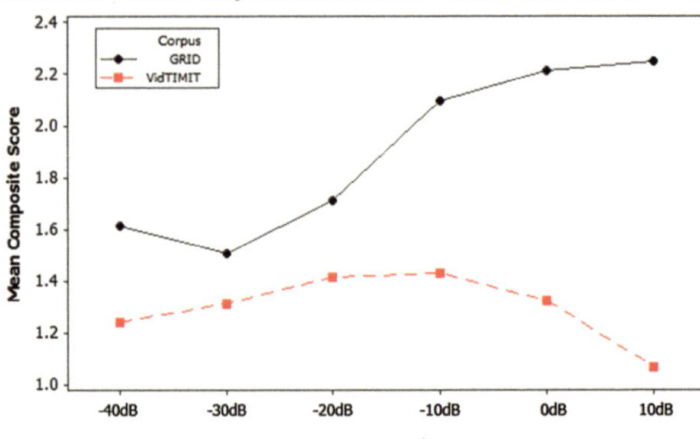

Fig. 5.12 Interaction plot for overall composite objective mean score of the GRID and VidTIMIT corpora at varying SNR levels

5.6 Inconsistent Audio Environment

5.6.1 Problem Description

Section 5.4 presented results that showed that when the two-stage system developed in this work was presented with a consistent noisy-speech environment, good results were found. Many audio-only source separation algorithms perform well in stable environments, but less strongly when the audiovisual environment is volatile and changeable. This section will demonstrate the effects of using a less consistent environment.

5.6.2 Experiment Setup

The system was set up in the same manner as described previously, with sentences from the GRID Corpus used in this experiment. The key difference from other experiments in this chapter is that rather than consistent aircraft noise being used as the noise source, speech sentences from the GRID Corpus are mixed with loud and inconsistent clapping at an SNR of –10 dB. The noise contains a period of silence at the start, and then the clapping is irregular and loud. Clean sentences from the corpus, with an example shown in Fig. 5.13, are mixed with this noise to produce the noisy-speech mixture. The resulting combined noisy-speech mixture is shown in Fig. 5.14. Firstly, the noisy mixture was processed using the standard audio-only beamformer. Secondly, this was then compared to filtering the same sentence with the two-stage audiovisual approach. In this section, a single example sentence was selected to be representative of all findings, as a very similar outcome was found

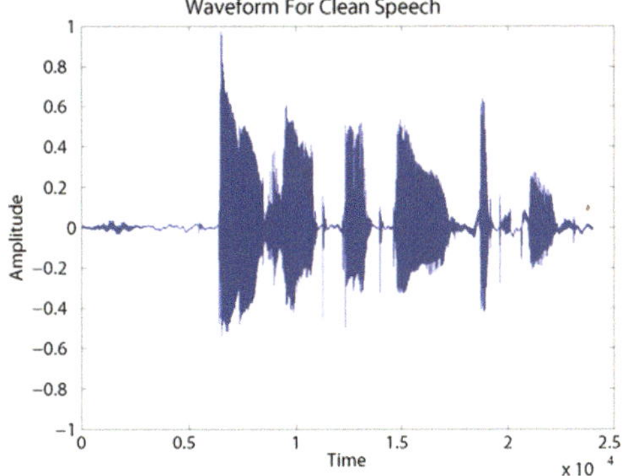

Fig. 5.13 Clean speech waveform for inconsistent noise environment experiment

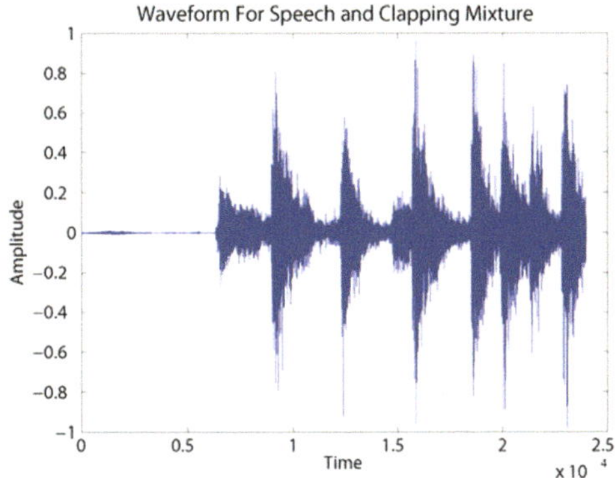

Fig. 5.14 Speech and noise mixture waveform for inconsistent noise environment experiment

for all speech sentences. To evaluate this experiment, it was felt that a visual comparison of results would be adequate for representation in this section because the difference between the two filtering approaches was found to be very clear for every test sentence tried.

5.6.3 Results and Discussion

After creating the mixture of speech and noise as described above and shown in Fig. 5.14, this was filtered using two different approaches. Firstly, the noisy-speech mixture was filtered using audio only beamforming. This was then compared to the same sentence, filtered using the two-stage audiovisual approach developed in this work.

An example of the results found with audio-only processing is shown in Fig. 5.15. It can be seen from a visual examination of the waveform that the sound completely fails to match the clean speech signal (Fig. 5.13). Listening to the file confirmed that the sound produced was not an accurate match for either clean-speech or noise. In

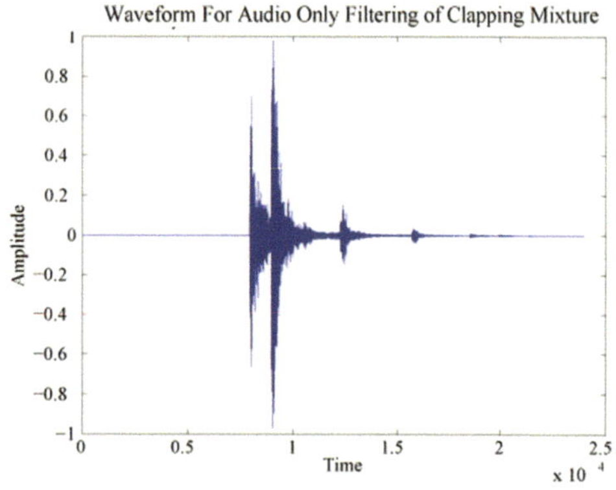

Fig. 5.15 Waveform for inconsistent noise environment experiment, result generated by audio only speech filtering

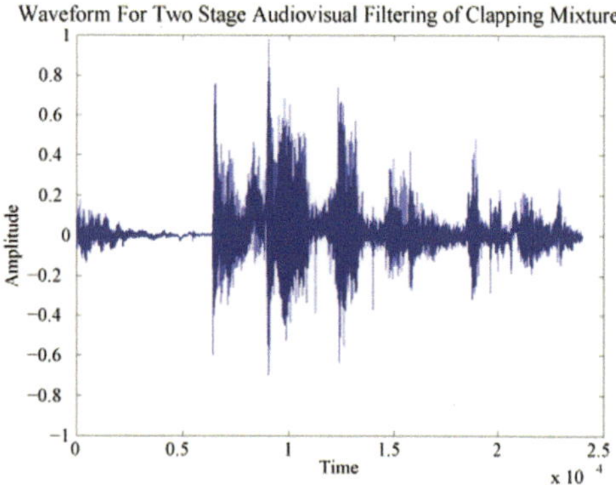

Fig. 5.16 Waveform for inconsistent noise environment experiment, result generated by two-stage audiovisual speech filtering

contrast to the simple audio-only processing (Fig. 5.15), when the additional pre-processing is used as part of the audiovisual filtering, then an improved result is produced. This can be seen in Fig. 5.16. The comparison shows that a much-improved sound is produced as a result of filtering, which when compared to the waveform Fig. 5.13, matches much more closely than using simple audio only filtering.

Overall, this section indicates that adding visually derived filtering may increase the flexibility of the system, as shown by the representative example result presented in this section. This shows that the addition of visually derived filtering can overcome limitations with single modality speech filtering, which can make it much more useful in environments where the noise source rapidly changes in SNR, volume, content, or when it can be difficult for audio only filtering to correctly distinguish between speech and noise.

References

1. A. Rix, J. Beerends, M. Hollier, A. Hekstra, Perceptual evaluation of speech quality (PESQ)—a new method for speech quality assessment of telephone networks and codecs, in *Proceedings of the 2001 IEEE International Conference on Acoustics, Speech, and Signal Processing, (ICASSP'01)*, vol. 2 (IEEE, 2001), pp. 749–752
2. Y. Hu, P. Loizou, Evaluation of objective measures for speech enhancement, in *Proceedings of the Interspeech* (Citeseer, 2006), pp. 1447–1450
3. J. Hansen, B. Pellom, An effective quality evaluation protocol for speech enhancement algorithms, in *ICSLP, Sydney, Australia*, (Citeseer, 1998), pp. 2819–2822
4. Y. Hu, P. Loizou, Evaluation of objective quality measures for speech enhancement. IEEE Trans. Audio, Speech Lang. Proc. **16**(1), 229–238 (2008)
5. A. Hussain, D. Campbell, Intelligibility improvements using binaural diverse sub-band processing applied to speech corrupted with automobile noise, in *IEE Proceedings—Vision, Image and Signal Processing*, vol. 148 (IET, 2001), pp. 127–132
6. I.-T. P.835, Subjective test methodology for evaluating speech communication systems that include noise suppression algorithm (2003)
7. I. Almajai, B. Milner, Enhancing audio speech using visual speech features, in *Proceedings of the Interspeech, Brighton, UK* (2009)
8. F. Fritsch, R. Carlson, Monotone piecewise cubic interpolation. SIAM J. Num. Anal. **17**(2), 238–246 (1980)
9. P.C. Loizou, *Speech Enhancement: Theory and Practice* (Signal Processing and Communications). 1 edn. (CRC 2007)
10. I.-T. P.862, Perceptual evaluation of speech quality (PESQ): an objective method for end-to-end speech quality assessment of narrow-band telephone networks and speech codecs (2001)
11. Malden Electronics Ltd., *Speech Quality Assessment Background Information for DSLA and MultiDSLA Users*. (Malden Electronics Ltd., 2004)
12. D. Klatt, Prediction of perceived phonetic distance from critical-band spectra: a first step, in *Acoustics, Speech, and Signal Processing, IEEE International Conference on ICASSP'82.*, vol. 7, (IEEE, 1982), pp. 1278–1281
13. P. Viola, M. Jones, Rapid object detection using a boosted cascade of simple features, in *IEEE Computer Society Conference on Computer Vision and Pattern Recognition*, vol. 1 (IEEE Computer Society 2001), pp. 511–518
14. A. Abel, A. Hussain, Q. Nguyen, F. Ringeval, M. Chetouani, M. Milgram, Maximising audio-visual correlation with automatic lip tracking and vowel based segmentation, in *Proceedings of the Biometric ID Management and Multimodal Communication: Joint COST 2101 and 2102 International Conference, BioID_MultiComm 2009, Madrid, Spain, 16–18 September 2009*, vol. 5707 (Springer 2009), pp. 65–72
15. I. Almajai, Audiovisual speech enhancement. Ph.D. thesis, University of East Anglia (2009)
16. B. Milner, I. Almajai, Noisy audio speech enhancement using Wiener filters derived from visual speech, in *Proceedings of the International Workshop on Auditory-Visual Speech Processing (AVSP)*
17. I. Almajai, B. Milner, Effective visually-derived Wiener filtering for audio-visual speech processing, in *Proceedings of the Interspeech, Brighton, UK* (2009)
18. Y. Lu, P. Loizou, A geometric approach to spectral subtraction. Speech Commun. **50**(6), 453–466 (2008)
19. A. Abel, A. Hussain, Novel two-stage audiovisual speech filtering in noisy environments. Cognit. Comput. 1–18 (2013)

Chapter 6
Towards Fuzzy Logic Based Multimodal Speech Filtering

Abstract After an investigation of state-of-the-art research in Chap. 3, Chap. 4 proposed a new two-stage audiovisual speech enhancement system that makes use of both audio and visual information to filter speech. The results of comprehensive testing in Chap. 5, identified a number of key strengths and weaknesses. It was concluded that although good results were found, and it demonstrated the feasibility of speech enhancement using visual information, there were also limitations. Chapter 5 concluded by identifying some potential refinements to the system, one of which was to extend the initial system with the use of fuzzy logic to make more cognitively inspired use of audio and visual speech information. This chapter presents a multimodal fuzzy logic based speech enhancement framework. Firstly, some limitations with the initial system are discussed. The decision to make use of a fuzzy logic based system is then justified. The chapter then presents a novel, multimodal, fuzzy logic based speech filtering framework. The utilisation of the audio and visual input data, the inputs to the fuzzy inference system, and the resulting fuzzy sets are described. The rules for the fuzzy logic based system, based on these fuzzy sets are then discussed. Finally, the challenges of thoroughly evaluating this initial system are then briefly discussed. While the work presented in this chapter does not represent a final and completed system, it is intended to demonstrate the feasibility of such an approach as an extension of the initial system presented previously in this work, showing that making more intelligent use of multimodal information is viable.

Keywords Multimodal · Fuzzy logic · Visual quality · Audio level · Speech processing

6.1 Limitations of Current Two-Stage System

The two-stage speech enhancement system presented previously was shown in Chap. 5 to be capable of producing positive results in environments where audio-only speech enhancement techniques were found to perform poorly. However, the results also identified a number of weaknesses with this two-stage approach, most significantly that when the SNR was relatively high, the two-stage approach

A. Abel and A. Hussain, *Cognitively Inspired Audiovisual Speech Filtering*, SpringerBriefs in Cognitive Computation, DOI 10.1007/978-3-319-13509-0_6

was outperformed by audio-only approaches due to the distortion introduced in the audiovisual filtering process. There were also limitations that apply more generally to visually aided speech filtering systems. These systems, whether visually derived Wiener filtering approaches [1] that estimate the noise-free audio signal with visual information, or multimodal beamforming systems [2] that use visual information for directional focus, rely on a clean source of visual information. The majority of multimodal speech research in the literature make the assumption that a good quality source of visual information is available at all times. The work presented previously makes the same initial assumption. This is acceptable for laboratory simulations, where a pre-recorded speech corpus is often used for research and development. However, in real world environments, such visual information is not always guaranteed to be present.

Visual speech information is particularly vulnerable to corruption. Consider a hypothetical scenario where a person is listening to a speaker. The listener may not always be looking directly at the speaker; their attention may be directed elsewhere at points during the conversation. There may be situations where the speaker turns their head, places a hand over their mouth, or another person walks between the speaker and listener, temporarily blocking the view of the speaker's face. The light level may also change, making it difficult to identify speech. These problems can affect both tracking and filtering, and in a real world scenario, have to be accounted for. There are also a number of other limitations that are not specific to visual information but apply to the audio domain. These were described in more depth in Chap. 3, but essentially there are many examples of certain types of speech filtering being vulnerable to environmental conditions. For example, directional microphones are often only recommended for wind-free environments.

The conclusion that can be drawn from this, is that there is no single specific speech processing algorithm that is guaranteed to perform strongly in all scenarios. Different approaches have their own weaknesses, so while visually derivedfiltering is vulnerable to missing visual data; beamforming is vulnerable to transient noise such as an unexpected loud handclap. Although the system presented in the previous chapters can offset these weaknesses to an extent, it introduces some of its own.

6.2 Fuzzy Logic Based Model Justification

To solve problems with real data, as discussed in Chap. 3, some commercially available audio-only hearing aids make use of decision rules to determine the extent and type of processing to apply to an input signal, based on various input detectors, and this allows for the adjustment of hearing aid settings to filter the input sound in a suitable manner. For example, as reported by [3], hearing aids exist that can take account of a number of detectors to analyse the input signal in order to classify the noise. Such an idea can also be seen in neuro-fuzzy systems [4]. Various audio input detectors can be used as an input to a set of decision rules, which then may apply different degrees of filtering, depending on the input.

6.2.1 *Requirements of Autonomous, Adaptive, and Context Aware Speech Filtering*

Therefore, it is proposed to extend the initial multimodal system to make use of audio and visual information in a more autonomous, adaptive and context aware manner, using cognitively inspired processing. Any such system has a number of requirements. Firstly, it has to be intelligent and context aware. By this, it is meant that the system takes account of the audio and visual speech environment and varies the processing decision accordingly. It also has to be adaptive and able to successfully process unpredictable environments.

Any proposed system should also be autonomous. As discussed in the review of directional microphone research in Chap. 3, [5, 6] reported that if the user was expected to manually determine the most suitable microphone setting (for example, omnidirectional or unidirectional) and switch their hearing aid to it when appropriate, then in the majority of cases, users would simply keep their hearing aids in omnidirectional mode. Another requirement of any state-of-the-art speech filtering system is that it should be capable of being tweaked and tuned without great difficulty. Hearing loss can vary widely between individuals, including hearing loss at specific frequencies or frequency ranges, and this wide variety of loss is not always handled well by existing hearing aid theory. Accordingly, when a modern programmable hearing aid is provided, patients are expected to undergo fitting sessions, where their hearing aid is programmed to better fit their individual hearing loss and comfort levels. Therefore, any proposed system should contain accessible parameters that can be tweaked and tailored in order to adapt to the hearing ability and preferences of the user.

Several alternative approaches were considered for use as part of this system, such as making use of ANNs, GMMs, HMMs, or a hybrid of these approaches, such as neuro-fuzzy approaches [4, 7]. Fuzzy logic was chosen for several reasons. Firstly, a fuzzy based system fulfils the requirement of being context aware, adaptive, and autonomous. Fuzzy logic is an approach that allows for uncertainty to be represented, therefore it is context aware, in that it is capable of responding to different changes in the environment, based on inputs into the system. It is also adaptive, in that it can respond to these inputs, so in the system presented in this chapter, the different inputs provide information about the environmental context (such as the level of noise), and the fuzzy-system makes a decision regarding the suitable processing choice, depending on this input. Fuzzy logic is also autonomous, and can make decisions without manual input. This can be seen in examples of other fuzzy inference systems, such as for controlling autonomous vehicles [8].

One key advantage to using fuzzy logic rather than equivalents such as ANNs and GMMs is that fuzzy logic is based on expert knowledge; this means that there are a number of rules that can be programmed, rather than requiring complete training, or the use of a mathematical model. This is of relevance, because a future requirement for a practical implementation of this system is that it should be possible to adjust settings and programs to suit individual users. For example, different users may have a different interpretation of what constitutes a 'very noisy environment', and so this can be customised in a fuzzy based system in a more straightforward manner than (for example) training ANNs or GMMs.

The preliminary system represented here makes use of very basic detectors, and could theoretically be represented using a different approach, such as with HMMs. However, it was decided that it was important to demonstrate a fuzzy-system, as it would arguably be more difficult to extend, train and implement a more sophisticated version of a HMM based system in future.

6.2.2 Fuzzy Logic Based Decision Making

Fuzzy logic was first proposed by [9], and allows uncertainty to be represented using the concept that a variable may belong to a set to an extent, but not completely. These sets are then used to create rules based on expert knowledge that can be evaluated to give an output based on uncertain input. Fuzzy inference is used in many applications, such as control of radio controlled vehicles [10] and noise cancellation [11]. It is important to clarify the distinction between fuzzy logic and probability. Fuzzy logic is not concerned with mathematically modelling a system (for example, HMMs [12]), and instead uses expert rules. While they both represent uncertainty, the semantics are different. For example, probability is based on the concept that a value has x probability of belonging to a set. This means that it may belong to the set, or it may not. On the other hand, an example of a fuzzy statement would say that a value belongs to a set to x extent. This is an important distinction. This is relevant because this concept can be applied to speech input, for example, audio input conditions can vary, depending on environmental conditions.

Fuzzy Sets
Traditionally, a crisp set represents values that either are members of a set or not. However, in fuzzy logic, values may partially belong to a set to a varying extent [13]. There can be several membership functions for an input, making it possible for an input to be a partial member of several sets simultaneously. So for example, a hypothetical input audio value can be part of a 'low noise' set, part of a 'high noise' set, or possibly a partial member of both sets.

Application of Fuzzy Operator
Fuzzy sets can then be used to perform a number of operations, including intersection (AND), unification (OR), and negation (NOT). This allows for the construction of rules. Rules make use of expert knowledge, and are expressed using variables

described by fuzzy sets. Each rule consists of two components, an antecedent and a consequent. The antecedent consists of the part of each rule before the 'Then' (e.g. IF a AND NOT b), and the consequent is the output (e.g. THEN c). These are used to create a series of rules. Each of the rules is evaluated to evaluate the extent to which it belongs to a chosen output. As a fuzzified input may belong to several sets, this means that several rules may be fired.

Aggregation
In order to produce an output value, each of the individual rule outputs is aggregated into a single fuzzy set. By this, it is meant that each output is evaluated, producing the output fuzzy values described above. Each rule is then combined to produce a single fuzzy set that encompasses all the rule outputs that were applicable.

Defuzzification
Defuzzification refers to the process of obtaining a single output value from the aggregated fuzzy set. There are several different methods, including centroid (the centre of the total aggregated area), middle of maximum (the average of the maximum value of the output set), and largest of maximum.

6.3 Potential Alternative Approaches

6.3.1 Hidden Markov Models

HMMs [12] are statistical Markov models, and are a type of temporal Bayesian network. While simpler Markov models have directly observable states, in a more complex hidden model, the states are hidden, and only the output values and their probabilities are directly observable. This approach is often used for time series data, where the data does not depend on previous time steps, but is theoretically only limited to that step. As described by [12], HMMs are defined by a set of states, an alphabet of changes between states, a transition probability matrix (giving probabilities of transitions from state to state), and an emission matrix (giving the probabilities of the outputs). There have been many different examples of making use of HMMs in the literature, with application to a range of problems, including speech recognition and enhancement [14, 15], robot control applications [16], multimodal emotion recognition [17], and many other tasks. A detailed overview of HMMs is provided by [12].

There are a number of general benefits to making use of HMMs. Firstly, they have a solid statistical grounding, with statisticians able to perform mathematical analysis of the results and manipulate the training process. Also, although hidden, there is also transparency, in that it is possible for the model to be read, so it is not a complete 'black box' solution. Finally, it is possible to incorporate prior knowledge into the model.

The most significant practical limitation is that there is a lack of potential flexibility in a speech filtering system designed using a statistical approach. While an initial system could feasibly be designed and trained, it would be designed for a very specific set of circumstances. As discussed elsewhere in this chapter, hearing aids are customised for individual users in fitting sessions. If using a HMM, a new model would have to be trained to suit, with the training data being labelled in order to match the comfort of an individual listener. This is a far more complex procedure than a domain expert (such as a trained audiologist) being able to tweak thresholds with a rule based model. Finally, there is the issue of training a model. Even a relatively simple approach with a limited number of states and outcomes will be expected to take account of a wide range of input data. This necessitates the acquisition of considerable quantities of carefully labelled data.

6.3.2 Neural Networks

Artificial Neural Networks (ANNs) [18–20] are considered to be a biologically inspired machine learning approach as they are theoretically similar to the structure of the brain (i.e. the biological connections between neurons). Generally, ANNs consist of input and output processing nodes, which are connected to a network using weighted connections. Neurons receive weighted values from incoming nodes, sum the received values, apply an appropriate activation function, and then pass the output to other nodes. The transfer function of an individual neuron refers to the threshold required before the neuron fires an output. There are many different types, including the commonly used logistic or tanh sigmoid function used commonly in Multilayer Perceptron (MLP) networks [21], which activates when the input exceeds a threshold level, and also LIF neurons, first proposed by [22], which are used in spiking networks [23]. There are many different topologies that are used, including Feedforward Networks [20] and recurrent neural networks [24, 25]. There have been many different utilisations of neural networks in the literature, such as classification problems [26], decision support [27], and speech filtering [4, 28].

ANNs are capable of solving complex problems, and have been used in the speech processing domain. It is potentially feasible for a neural network to be developed to make a decision regarding the most suitable processing method when presented with novel information.

However, there are some issues with using this approach. Like the HMM method discussed previously, there is the fundamental issue of customising a neural network based approach for individual listener comfort. The same issue applies to many other machine learning approaches such as support vector machines and GMMs. Although they could feasibly solve the particular problem discussed in this chapter, their general purpose utilisation in a future system becomes more problematic. Neural network approaches also have the issue of being effectively a 'black box' system. Tweaking and refining is not a simple matter.

6.4 Fuzzy Based Multimodal Speech Enhancement Framework

In light of the limitations that have been outlined earlier in this chapter, the initial multimodal two-stage speech enhancement system presented in Chap. 4 has been extended to become more sophisticated by developing a fuzzy logic based system. It was felt that using fuzzy rules represented the most practical solution, and could theoretically be implemented and modified in future designs of hearing aids. To do this, a fuzzy logic controller has been implemented to determine the most suitable method for processing an individual frame of speech.

6.4.1 Overall Design Framework of Fuzzy System

The initial system shown in Fig. 4.1 is extended further by the integration of a fuzzy logic controller and the subsequent adjustment of the speech filtering options. The basic components introduced in Chap. 4 are unchanged. Visual tracking and feature extraction is handled in the same manner, as is the audio feature extraction process. With regard to speech processing, the two processing options used, visually derived Wiener filtering and audio-only beamforming, remain unchanged. However, the difference is that one or both of these stages may be bypassed on a frame-by-frame basis, depending on the inputs received by the fuzzy logic controller. This redesigned framework is shown in Fig. 6.1.

Figure 6.1 shows the high level extended system diagram with the alternative speech processing options. Depending on the inputs to the fuzzy logic controller, the type of processing performed on the input signal may vary from frame-to-frame. So for example, if it is detected that there is very little audio activity in a particular frame, then it may be decided to leave that frame unfiltered. Alternatively, if a moderate amount of audio energy is detected, then it may be decided that audio-only beamforming is the most appropriate processing method. If however, a lot of audio

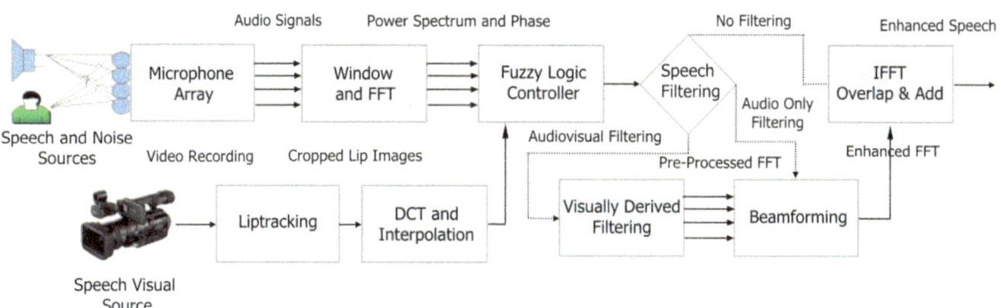

Fig. 6.1 System diagram of proposed fuzzy logic based two-stage multimodal speech enhancement system. This is an extension of Fig. 4.1, with the addition of a fuzzy logic controller to receive inputs and decide suitable processing options on a frame-by-frame basis

activity is detected in a particular frame and the visual information is considered to be of good quality, then the full two-stage process may be used.

The decision as to which option is to be used is taken with the aid of a number of detectors applied to the input signal. In the initial implementation presented in this chapter, three detectors are used. An audio level detector, a visual quality detector, and a feedback input of the processing decision made in the previous frame. Fuzzy logic rules are then used to determine the most suitable processing method, depending on the input. Each individual frame is processed in the manner judged by the fuzzy logic controller to be most suitable.

6.4.2 Fuzzy Logic Based Framework Inputs

As discussed in Sect. 6.2, each input variable must be decomposed into fuzzy sets, consisting of a number of membership functions. The composition of these membership functions can vary in size and shape, based on the preference of the designer [29], and for the work in this book, it was decided to make use of trapezoid membership functions for all inputs in order to ensure consistency. The fuzzy-system diagram is shown in Fig. 6.2, and it can be seen that there are three inputs to consider, audio level, visual quality, and previous frame processing decision.

6.4.2.1 Visual Quality Fuzzy Input Variable

The first input variable is the visual quality. This measures the level of detail found in each cropped ROI. As the system is audiovisual, visual information is a key component of the processing. However, this information can be of varying quality. There are occasions when the entire lip region is visible, but there are also occasions

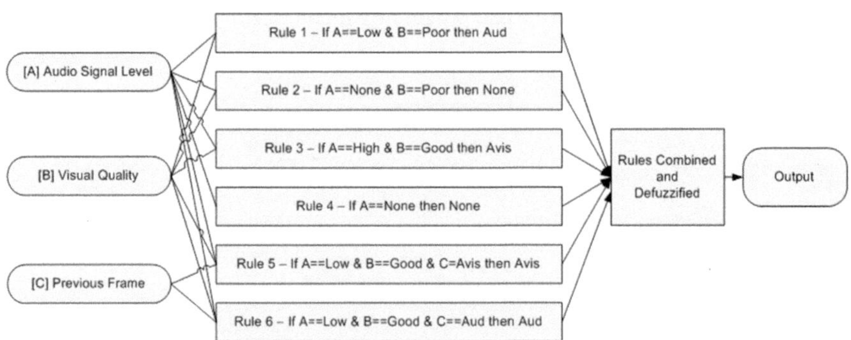

Fig. 6.2 Diagram of fuzzy logic components, showing the three chosen fuzzy inputs and the list of rules to be applied

when the lip-tracker returns an incorrect result due to scenarios such as the speaker turning their head. There are also occasions when the lip region may be blurred due to movement, or only a partial ROI is returned. One potential approach was to make use of a machine learning technique such as a HMM to create a model to evaluate the ROI and return a score to use as a fuzzy input variable. However, it was felt that this was not required for the initial implementation. Instead, a simpler approach was devised that made use of the input DCT vector.

In order to determine the most suitable input variable, a custom corpus was recorded using real data from a variety of volunteers. This is discussed in more depth in Sect. 7.3, and a number of trial videos were evaluated to calculate the most suitable value, with various variables investigated, such as the DCT input vector, and the tracker parameters of the actual cropped images. This identified that the fourth DCT coefficient was consistently a better representation of the accuracy of the cropped DCT than any other single factor, and so this was used to create a mean value. As the DCT transform represents pixel intensity, it was calculated that while the value of this would vary from image to image, the fourth coefficient value would remain relatively consistent. Therefore, for each frame, the absolute value of the fourth DCT coefficient was calculated. This was then compared to a moving average of up to the 10 previous frames that were considered to also be of good quality, and the difference between this moving average and the coefficient represented the visual input variable.

To create this moving average, one assumption was made, that the first value of each sentence was successfully identified with the Viola-Jones detector [30]. This first value was used as the initial moving average mean value. For the second frame onwards, the new value was compared to the mean of the moving average. If the new value was considered to be within a threshold (preliminary trials identified an appropriate threshold to be 2000), then this value was considered to be suitable, and so was added to the moving average. To take account of variations in speech from frame-to-frame, only a maximum of the 10 most recent values were considered as part of the moving average. This moving average threshold aims to minimise incidences of incorrect results being added to the moving average.

Preliminary trials found that examples of poor quality visual information tended to result in a greater difference from the mean than good information, and so this approach was found to be suitable. The trapezoidal membership functions are shown in Fig. 6.3.

Figure 6.3 shows that there are two membership functions, 'Good' and 'Poor'. The lower the input value, the closer to the mean and therefore the better the frame of visual data was considered to be. However, as values for individual speakers could vary depending on factors such as the size of the detected ROI and the degree of emotion in their speech (for example, affecting the size of mouth opening), there was no fixed value that was guaranteed to work for every speaker, and therefore a crisp set was not considered to be suitable. As a fuzzy membership function was used, it was considered that a visual quality value of less than 800 was definitely an example of a good frame of visual information. Between 800 and 2000, then it could be sometimes considered a partial member of the good set in that there was some

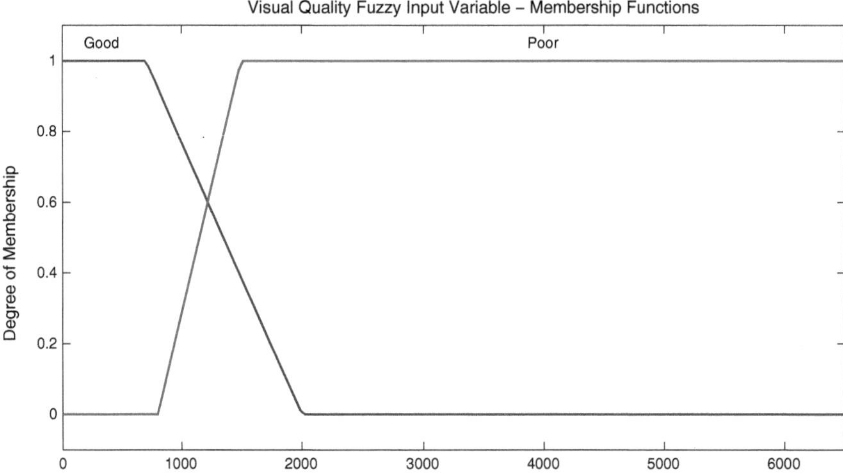

Fig. 6.3 Switching logic input parameter: visual detail level. Depending on the level of visual detail, the estimated parameter can be considered to be 'Good' or 'Poor' to varying extents

ambiguity depending on the speaker, and also there were examples of partial frames (where only part of the ROI was accurate). This justified the decision to use fuzzy input variables.

6.4.2.2 Audio Power Fuzzy Input Variable

The second input variable to be used is the audio power level. This considers how much acoustic activity there is in an individual frame of speech. The membership functions are shown in Fig. 6.4. This variable does not consider the problem of voice activity detection, and so does not attempt to distinguish between speech and noise. One reason for this is that the system is designed to be tested in extremely noisy environments, and audio-only VAD techniques do not always perform well in these environments. As shown in the results in Chap. 5, at an extremely low SNR, no speech at all can be identified in noisy input speech mixtures. It is possible to devise an audiovisual VAD [31], and this could represent future potential development.

To calculate the audio power in each input speech frame, the frame is first converted back to the time domain to return the amplitude waveform for that frame of speech. The mean of the absolute values of the frame is then found. This represents the level of the audio power. The fuzzy set that the audio power input variable belongs to is then calculated based on this input, as shown in Fig. 6.4. To take account of extremely noisy input variables, due to the extremely low SNR that the system is tested with, the largest trapezoidal membership function is the 'High' value, which has a maximum value of 25. Figure 6.4 shows only the fuzzy membership functions for values less than 1.5.

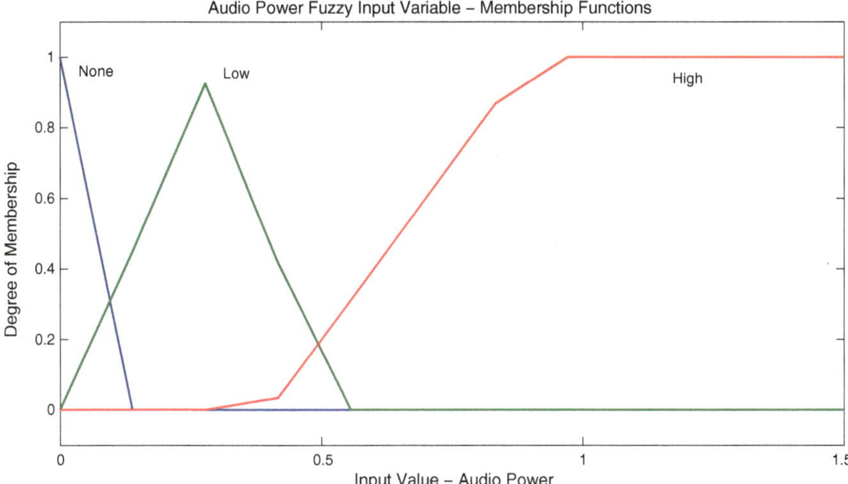

Fig. 6.4 Switching logic input parameter: audio frame power, showing only membership functions for values ranging from 0 to 1.5. Depending on the level of audio power, the estimated parameter can be considered to be 'None', 'Low', or 'High'

Figure 6.4 shows that if the level is recorded as being very low (less than 0.015), the input level is considered to belong to the 'None' membership function. However, as the level detector is very sensitive, it can be seen that any positive level (ranging from 0.009 to 0.5) is also part of the 'Low' fuzzy-set to an extent. Finally, any values greater than 0.4 were considered to be a member of the 'High' set to an extent, and values greater than 0.9 were considered to fully belong to the 'High' set. These values were set by using trial data.

6.4.2.3 Previous Frame Fuzzy Input Variable

The third input variable is the previous frame fuzzy logic output. This is simply a feedback variable that passes in the fuzzy logic controller output from the previous frame. As stated, there are three different processing options, and this can be seen in Fig. 6.5, which is valid for the representation of both the controller output and the third input. The reason for this third input is to act as a smoothing function in marginal cases. For example, the audio and visual inputs may produce input variables that lie near the thresholds between two possible processing options. Small changes in subsequent frames may produce a radically different processing decision from frame-to-frame. As a consequence, the output sound quality may be of poor listening comfort (as is sometimes found in conventional hearing aids when the engaging/adaption/attack configuration is set poorly, resulting in a 'choppy' sound, as discussed by [32]). The use of the previous frame in marginal scenarios is designed to limit this. This input performs the role of engaging/adaption/attack configuration in

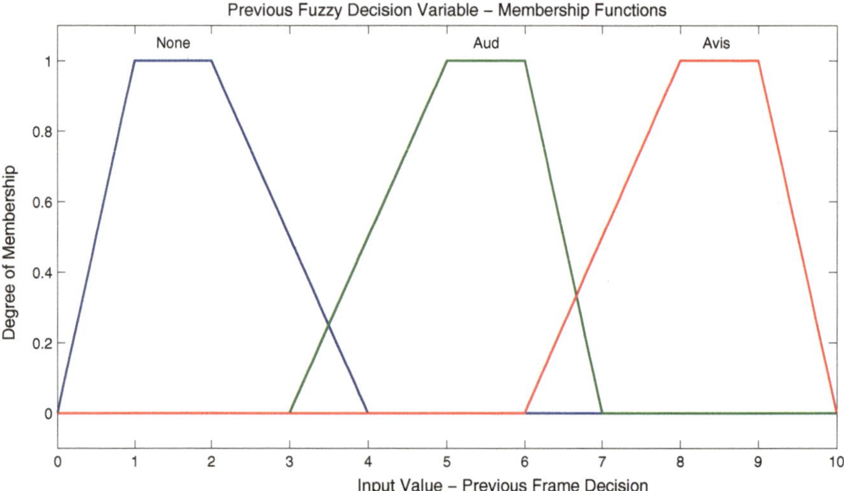

Fig. 6.5 Switching logic input parameter: previous frame output. This input variable considers the processing method chosen in the previous frame. Therefore, this input fuzzy set diagram matches the output choice. This input is useful in marginal examples, as will be discussed later in this chapter

this preliminary system, as it introduces what is effectively a small delay into processing changes. An evaluation of the role of this input variable on the output decision is discussed in Chap. 7, demonstrating that this input variable plays an important role in limiting changes in the fuzzy output. Figure 6.5 shows the three trapezoidal membership functions representing the possible processing choices, 'None' (leave the frame unprocessed), 'Aud' (audio-only beamforming), and 'Avis' (meaning to use the audiovisual approach).

6.4.3 Fuzzy Logic Based Switching Supervisor

6.4.3.1 Fuzzy Rules for Switching Decision

The fuzzy logic controller is used to determine the most suitable speech processing method to apply to an individual frame of speech, based on the fuzzy input variables defined in the previous section. One difference between simple rules with crisp sets and fuzzy-based rules is that the rules are fired to varying degrees, depending on the extent to which the input variables are part of potentially overlapping membership functions. If an input variable is part of more than one membership function (for example, the audio level may be considered to be part of both the 'None' and 'Low' sets, then contrasting rules may be fired, with the strength of each rule depending on the extent to which the input variable is part of the relevant fuzzy set. These rule outputs are aggregated to produce one fuzzy output set encompassing all of the rules that were fired. Finally, this is defuzzified (again, as described in Sect. 6.2) to produce one single fuzzy output decision value. In this work, the centroid value was used.

The input variables to the fuzzy-system are described above and are the audio level (audSigPow), visual quality (visQuality), and the previous frame controller output (prevFrame). An input variable may simultaneously belong to more than one fuzzy set to varying extents. The processing output options are no processing (a), audio-only processing (b), or two-stage audiovisual processing (c). The complete set of rules used in this system is listed as follows:

- Rule 1: IF audioSigPow IS low AND visQuality IS poor THEN process is b
- Rule 2: IF audioSigPow IS none AND visQuality IS poor THEN process is a
- Rule 3: IF audioSigPow IS high AND visQuality IS good THEN process is c
- Rule 4: IF audioSigPow IS none THEN process is a
- Rule 5: IF audioSigPow IS low AND visQuality IS Good AND prevFrame IS avis THEN process is c
- Rule 6: IF audioSigPow IS low AND visQuality IS Good AND prevFrame IS aud THEN process is b

Rule 1 activates audio-only processing if the audio input variable belongs to the 'Low' fuzzy set and the visual quality is defined as being 'Poor'. Rules 2 and 4 ensure that the frame is left unfiltered if the audio level is found to be so low that the audio level is defined as being 'None'. Rule 3 activates audiovisual processing if there is a sufficient level of noise, and if visual information of an adequate quality is available. Rules 5 and 6 are designed to take effect in scenarios where the potential choice of processing algorithm is ambiguous. If the audio level is defined as 'Low', but 'Good' quality visual information is available, then the previous frame input is also considered. Rule 5 activates audiovisual processing if the previous frame output was also audiovisual, and rule 6 activates audio-only processing if the previous frame decision was audio-only. This is intended to ensure continuity between frames and prevent rapid frame-by-frame changes that act as an irritant to listeners.

6.4.3.2 Fuzzy Inference Procedures for Switching Logic

For each frame of speech, the three input variables are calculated and input to the fuzzy rules. For brevity, we focus on the example of most interest, if a frame of speech is in a marginal area where it could be either audio-only, or audiovisual processing used.Both examples, shown in Figs. 6.6 and 6.7 have an audio input variable which is defined as being 0.454. This is a value which belongs to two potential fuzzy sets. The visual quality level is set to 847 in both examples, meaning that this visual information is considered to be a member of both the 'Good' and 'Poor' membership functions. As the examples show, both rules 1 and 3 are fired, and both have a similar level of dominance when it comes to establishing the output. These values mean that small changes in the audio variable in successive frames may result in an entirely different

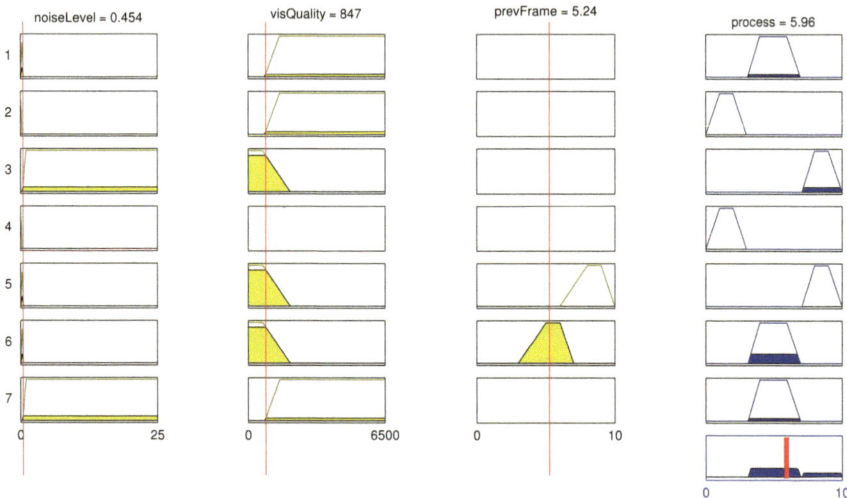

Fig. 6.6 Demonstration of rule selection for a marginal example. Good quality visual information is available, and the audio level is measured as being applicable to both moderate and high levels. In this case, the previous frame output information is used, which in this example was audio-only processing, and so the audio-only option is chosen again

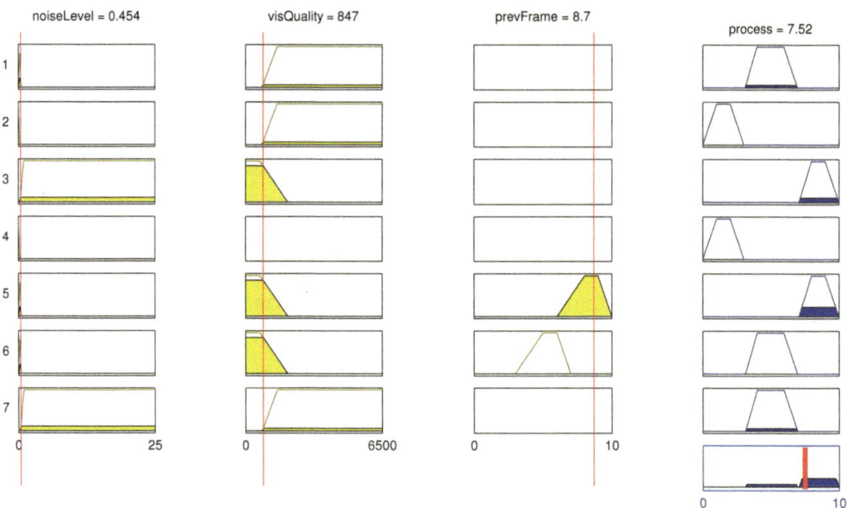

Fig. 6.7 Demonstration of rule selection for a marginal example. Good quality visual information is available, and the audio level is measured as being applicable to both moderate and high levels. In this case, the previous frame output information is used, which in this example was audiovisual processing, and so the audiovisual option is chosen again in this frame

type of processing being used from frame-to-frame, despite there being potentially only a very small change in environmental conditions. This is undesirable because to the listener, rapid and unnecessary switching between processing methods can often result in an unpleasant listening experience. As can be seen in Fig. 6.6, although rules 1 and 3 are fired, rule 6 is also fired. Rule 6 considers the previous frame, and as this is defined in this example as using audio-only processing, this becomes dominant, and so when defuzzification takes place, audio-only processing is chosen. Figure 6.7 uses the same audio and visual variables, but with the previous frame input variable being defined as audiovisual. In this case, the audiovisual technique is chosen by the fuzzy logic controller. A more detailed evaluation of this system is presented in the next chapter.

References

1. B. Milner, I. Almajai, Noisy audio speech enhancement using Wiener filters derived from visual speech, in *Proceedings of the International Workshop on Auditory-Visual Speech Processing (AVSP)*
2. B. Rivet, L. Girin, C. Serviere, D.-T. Pham, C. Jutten, Using a visual voice activity detector to regularize the permutations in blind separation of convolutive speech mixtures, in *15th International Conference on Digital Signal Processing* (2007), pp. 223–226
3. N. Tellier, H. Arndt, H. Luo, Speech or noise? Using signal detection and noise reduction. Hear. Rev. **10**(6), 48–51 (2003)
4. A. Esposito, E. Ezin, C. Reyes-Garcia, Designing a fast neuro-fuzzy system for speech noise cancellation, in *MICAI 2000: Advances in Artificial Intelligence* (2000), pp. 482–492
5. M. Cord, R. Surr, B. Walden, L. Olson, Performance of directional microphone hearing aids in everyday life. J. Am. Acad. Audiol. **13**(6), 295–307 (2002)
6. M. Cord, R. Surr, B. Walden, O. Dyrlund, Relationship between laboratory measures of directional advantage and everyday success with directional microphone hearing aids. J. Am. Acad. Audiol. **15**(5), 353–364 (2004)
7. M. El-Wakdy, E. El-Sehely, M. El-Tokhy, A. El-Hennawy, N. Mastorakis, V. Mladenov, Z. Bojkovic, D. Simian, S. Kartalopoulos, A. Varonides, et al., Speech recognition using a wavelet transform to establish fuzzy inference system through subtractive clustering and neural network (ANFIS), in *Proceedings of the International Conference on Mathematics and Computers in Science and Engineering*, vol. 12 (WSEAS, 2008)
8. R.A. Abdullah, A. Hussain, M.M. Polycarpou, Fuzzy logic based switching and tuning supervisor for a multi-variable multiple controller, in *IEEE International Fuzzy Systems Conference, FUZZ-IEEE 2007* (IEEE, 2007), pp. 1–6
9. L. Zadeh, Fuzzy sets*. Inf. Control **8**(3), 338–353 (1965)
10. K. Tanaka, M. Iwasaki, H. Wang, Switching control of an R/C hovercraft: stabilization and smooth switching. IEEE Trans. Syst. Man Cybern. Part B: Cybern. **31**(6), 853–863 (2001)
11. C. Chang, K. Shyu, A self-tuning fuzzy filtered-U algorithm for the application of active noise cancellation. IEEE Trans. Circuits Syst. I: Fundam. Theory Appl. **49**(9), 1325–1333 (2002)
12. Z. Ghahramani, An introduction to hidden Markov models and Bayesian networks. Int. J. Pattern Recognit. Artif. Intell. **15**(01), 9–42 (2001)
13. M. Hellmann, *Fuzzy Logic Introduction* (Université de Rennes, France, 2001)
14. P. Bansal, A. Kant, S. Kumar, A. Sharda, S. Gupta, Improved hybrid model of HMM/GMM for speech recognition. Intell. Technol. Appl. 69–74 (2008)
15. J. Hershey, M. Casey, Audio-visual sound separation via hidden Markov models. Adv. Neural Inf. Process. Syst. **14**, 1173–1180 (2001)

16. I. El-emary, M. Fezari, H. Attoui, Hidden markov model/gaussian mixture models (HMM/GMM) based voice command system: a way to improve the control of remotely operated robot arm TR45. Sci. Res. Essays **6**(2), 341–350 (2011)
17. Z. Zeng, J. Tu, B. Pianfetti, M. Liu, T. Zhang, Z. Zhang, T. Huang, S. Levinson, Audio-visual affect recognition through multi-stream fused HMM for HCI, in *IEEE Computer Society Conference on Computer Vision and Pattern Recognition. CVPR*, vol. 2 (IEEE, 2005), pp. 967–972
18. J. Zurada, *Introduction to Artificial Neural Systems* (West St. Paul, 1992)
19. A. Zayed, A. Hussain, R. Abdullah, A novel multiple-controller incorporating a radial basis function neural network based generalized learning model. Neurocomputing **69**(16), 1868–1881 (2006)
20. S. Haykin, *Neural Networks: A Comprehensive Foundation* (Prentice Hall, Upper Saddle River, 1998)
21. D. Rumelhart, G. Hintont, R. Williams, Learning representations by back-propagating errors. Nature **323**(6088), 533–536 (1986)
22. E. Adrian, *The Basis of Sensation* (W.W. Norton & Co., New York, 1928)
23. W. Maass, Networks of spiking neurons: the third generation of neural network models. Neural Netw. **10**(9), 1659–1671 (1997)
24. H. Jaeger, The "echo state" approach to analysing and training recurrent neural networks-with an erratum note. Tecnical report GMD report, vol. 148 (2001)
25. W. Maass, T. Natschläger, H. Markram, Real-time computing without stable states: a new framework for neural computation based on perturbations. Neural Comput. **14**(11), 2531–2560 (2002)
26. M. Newton, L. Smith, A neurally inspired musical instrument classification system based upon the sound onset. J. Acoust. Soc. Am. **6**, 4785–4798 (2012)
27. P. Lisboa, A. Taktak, The use of artificial neural networks in decision support in cancer: a systematic review. Neural Netw. **19**(4), 408–415 (2006)
28. A. Hussain, D. Campbell, Binaural sub-band adaptive speech enhancement using artificial neural networks. Speech Commun. **25**(1), 177–186 (1998)
29. A. Bagis, Determining fuzzy membership functions with tabu search—an application to control. Fuzzy Sets Syst. **139**(1), 209–225 (2003)
30. P. Viola, M. Jones, Rapid object detection using a boosted cascade of simple features, in *IEEE Computer Society Conference on Computer Vision and Pattern Recognition*, vol. 1 (IEEE Computer Society, 2001), pp. 511–518
31. I. Almajai, B. Milner, Effective visually-derived Wiener filtering for audio-visual speech processing, in *Proceedingf of the Interspeech, Brighton, UK* (2009)
32. K. Chung, Challenges and recent developments in hearing aids. Part I. Speech understanding in noise, microphone technologies and noise reduction algorithms. Trends Amplif. **8**(3), 83–124 (2004)

Chapter 7
Evaluation of Fuzzy Logic Proof of Concept

Abstract This book presented a novel two-stage speech enhancement system in Chap. 4, which was then thoroughly tested in Chap. 5. As a result of these tests, although promising results were found, some limitations with both this system and speech enhancement systems generally were identified, including limitations with the visually derived filtering utilised in this research. This resulted in the development of a proposed fuzzy logic based two-stage speech processing system that uses fuzzy input variables to determine the most appropriate method of processing an individual frame of speech. This preliminary system was described in depth in Chap. 6. This chapter presents an evaluation of this preliminary system, firstly discussing the need for testing, and the requirements for testing a system that is designed to deal with more realistic speech processing scenarios. To present a full description of the limitations with the system in its current preliminary state, and to better meet the requirements of more realistic test scenarios, a new corpus is used, recorded specifically to provide challenging audiovisual data. This corpus is then used as part of a series of challenging experiments in order to evaluate the performance of this system.

Keywords Multimodal · Fuzzy logic · Speech filtering · Audio input · Visual input · Cognitively inspired · Results

7.1 Testing Requirements

To test the system presented in Chap. 6, it is not possible to simply use sentences from the GRID corpus, as it does not contain the variation in audio and visual data quality required to be representative of data that a system is likely to be expected to process successfully, and so more realistic audiovisual data is required for valid testing. This means that the audio data should ideally contain noise of different levels, with the speaker then adjusting their speech correspondingly as they would in a real situation. Conversational scenarios should also be of longer length than simple two or three second sentences, with examples of overlapping speakers, silences, turn-taking, and emotional speech. The visual data should also be of variable quality, with examples of speakers moving in a free and natural manner.

A. Abel and A. Hussain, *Cognitively Inspired Audiovisual Speech Filtering*,
SpringerBriefs in Cognitive Computation, DOI 10.1007/978-3-319-13509-0_7

7.2 Experimentation Limitations

Given the current preliminary implementation of the system, it is extremely challenging to fully evaluate the system. With the fuzzy logic based system presented in this chapter still in a preliminary stage of development, output filtered speech results using real data should be interpreted with a degree of caution. Some prior examples of audiovisual data tests in the literature include work by [1], who train and test with the same single speaker corpus using consistent broadband noise at a relatively high SNR, and audiovisual source separation by [2]. Although [2] make use of more realistic data, they make a number of assumptions regarding room size and availability of visual information. The prior work in the literature demonstrates that audiovisual speech filtering systems are often tested with fairly limited parameters.

In addition, due to the preliminary state of implementation of the system, noise had to be added separately in the simulated room so that the impulse responses could be calculated, which is a requirement for beamforming to be successfully performed. Recording of noisy data by adding a noise source was attempted, but as expected, the system was not able to successfully and reliably process this data, and so this data was not used in the final testing. Another related issue concerned data synchronisation. To use multiple microphones, the data has to be synchronised. If the microphones and camera are recorded individually, synchronising them by hand (matching the audio and visual data by inspection and adjustment) can prove to be extremely time consuming, and so a full hardware multi-microphone corpus was not pursued here. In addition, there are limitations with the visually derived filtering. It was found that when the system was tested with novel speakers, it did not generalise well, and so when tested with real data, it is expected to perform poorly. Another issue is that the simulated room environment (with mixtures of speech and noise sources) was designed explicitly to demonstrate the effectiveness of beamforming, and so tests using this scenario provide an artificial benefit to using beamforming that would not be expected in a real environment (a similar issue to the improvement in results with using directional microphones in laboratory environments rather than in real environments). Overall, it is understandable why other similar speech enhancement work such as by [3, 4] makes use of very limited experimental scenarios.

7.3 Recording of Challenging Audiovisual Speech Corpus

7.3.1 Corpus Configuration

Volunteers were asked to perform two tasks. Firstly, a reading task, where they read either a short story or a news article. For this task, they were recorded reading for a minute in a quiet environment, and then a minute in an environment with a variable level of noise (a mix of music tracks, with the volume varied randomly). This allowed for both good quality audio data, and also poorer quality raw data to be collected (with the Lombard Effect having an impact on resulting visual data).

The second scenario was a conversational task, where volunteers were encouraged to speak in a more natural manner. Volunteers were recorded in pairs at a table facing each other, with one speaker recorded at a time. By this it is meant that while the speakers were facing each other and making conversation, the camera was only pointed at one speaker. This allowed more natural and relaxed speech, and the volunteers were also told that they were allowed to move freely and did not have to look directly into the camera at all times. Volunteers were given a choice of topics to choose from, such as a TV programme, something that interested them in the media, or could choose their own conversation topic. Due to this being a conversation rather than continuous speech from a single recorded speaker, there were occasional silences, or speech from the other participant. Again, each speaker was asked to speak for one minute in a quiet environment, and one minute in a noisy environment, although the noisy data was subsequently not used in this work.

7.3.1.1 Finalised Corpus Description

To record volunteers carrying out the tasks described above, a single camera (a Microsoft VX2000 Lifecam) was used with an integrated microphone to record speech in a quiet room. The visual data was recorded at a resolution of 640×480. However, there were some issues with the recording process. Firstly, the video camera had automatic brightness adjustment enabled, and so a small number of frames were considerably darker due to occasional automatic readjustment. An example of this can be seen in the lower image in Fig. 7.1. There were also a number of glitches in the recording. An example of this can be seen in the top image in Fig. 7.1. In this image, the camera has not recorded the head of the speaker in a single frame, although in subsequent and preceding frames, the head is not missing.

The final corpus contained data from eight speakers, four male, four female. Six of the eight speakers spoke English (five with a Scottish accent and one English), and two were recorded speaking Bulgarian. For each speaker, four minutes of raw data were theoretically available, one minute of quiet conversation, one minute of variable noisy conversation, and then one minute each of noisy and quiet reading. Some example frames of the recorded volunteers are shown in Fig. 7.2. As part of the requirement for the visual data to be challenging and of variable quality, speakers were expected to move naturally. This led to variable quality visual data, with some examples shown in Fig. 7.3. The top image shows an example of the speaker in the process of moving their hand in front of their mouth, meaning that lip information is not available. The lower image shows an example of the speaker turning their head to one side.

Of the initial 32 min of raw data, approximately 6 min was unavailable due to the recording problems described earlier. The data was divided into 20 s clips because of processing and testing requirements. This sentence length was significantly longer than available in the pre-recorded corpora, and was felt to be long enough to test the operation of the fuzzy-system, while still being short enough to process relatively efficiently. A number of these 20 s clips were then chosen for use as part of the testing

Fig. 7.1 Examples of poor quality visual data due to issues with recording. The *top image* shows an example of a glitch during recording, resulting in the face region being removed. The *bottom image* shows a situation where light conditions have changed, resulting in a temporarily darker image

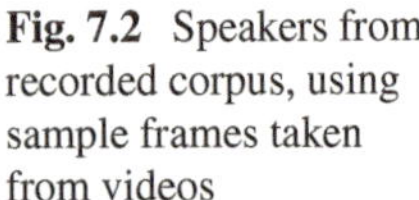

Fig. 7.2 Speakers from recorded corpus, using sample frames taken from videos

Fig. 7.3 Examples of poor quality visual data due to speaker actions. The *top image* shows a frame where the speaker has their hand over their mouth due to gesturing during emotional speech. The *bottom speaker* is in the process of quickly turning their head, and as a result the mouth region is partially obscured, and the face is blurred

process. These were chosen to represent a mixture of different conditions and data quality. As mentioned previously, only the sentences without noise were used.

7.4 Fuzzy Input Variable Evaluation

The previous chapter presented a description of a fuzzy logic based speech processing system, which made use of three fuzzy input variables, the audio power within a frame, visual data quality, and also the output decision of the previous frame, which is fed back in as an input. While the audio input variable is very closely related to the audio signal, the effectiveness of the other two input variables is of great interest.

7.4.1 Visual Quality Fuzzy Indicator

7.4.1.1 Problem Description

As described in Sect. 6.4, one input variable used in the system was the visual quality variable. There was an assumption made that the initial ROI was accurately detected, and subsequent frames were calculated in terms of the difference from the mean of the absolute value of the fourth DCT coefficient. To take account of natural movement over time, a moving average of the previous 10 frames was used, with only frames that were considered to be within a threshold added to the moving average. This value was then used as the visual fuzzy input value.

7.4.1.2 Experiment Setup

20 sentences from the corpus described in Sect. 7.3 were used for evaluation. This included 10 sentences recorded in a quiet environment, and 10 recorded in an environment with some noise present. In addition, to ensure that a range of different visual challenges was represented, 10 reading examples, and 10 conversation examples were used, from a number of different speakers. This ensured that challenging data was used and provided a rigorous test of this fuzzy input variable.

For each sentence, a manual review of each cropped lip image was performed. This involved inspecting each frame and assigning it a value. A frame that was considered to be of good quality (in that it showed the whole lip-region) was given a score of 1. An image that was considered to be of lower quality (either showing only part of the lip-region or the wrong region) was given a score of 2. Finally, an extremely poor result (one where no ROI at all was identified) was given a score of 3.

The manual input estimation of every frame of each sentence was compared to the equivalent fuzzy input variable. As the variable can vary in value between 0 and 6000+, with a lower value indicating less difference from the mean, then based on preliminary trials, a value of less than 1000 was given a score of 1 (some examples of this are shown in Fig. 7.4), a value of less than 4500 but greater than 1000 was given a score of 2 (as shown by the examples in Fig. 7.5), and anything greater than 4500 was given a score of 3, representing examples where no ROI was identified, as shown in Fig. 7.6. This allowed the visual input variable output to be mapped to the manual estimation.

7.4.1.3 Summary of Results

Firstly, when taking 20 all sentences into account (whether recorded in a quiet or noisy environment, or as part of a reading or conversation task), after interpolation there were a total of 39,975 frames of data. Of these, 92.15 % produced a correct

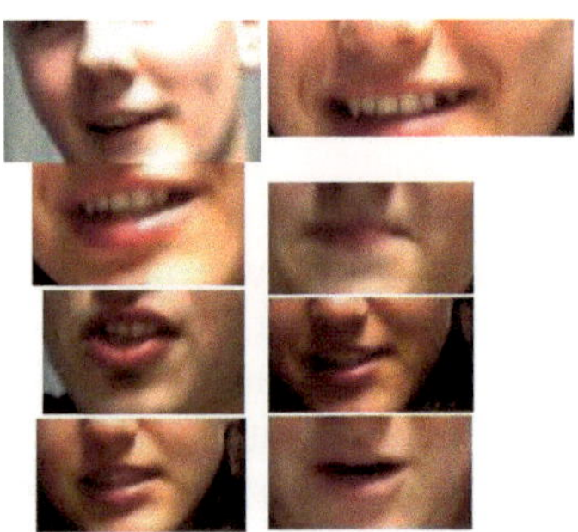

Fig. 7.4 Examples of lip images regarded to be successfully detected. It can be seen that the images are of varying dimensionality, and also include varying levels of additional facial detail depending on the results of the Viola-Jones lip detector

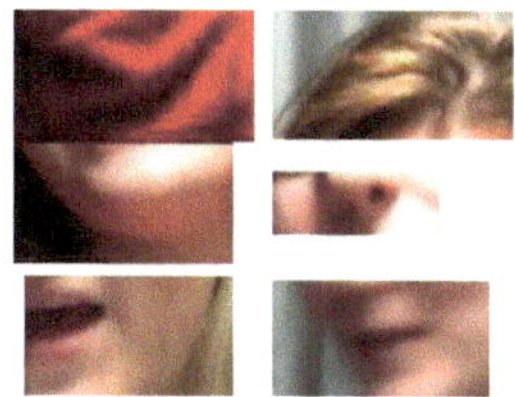

Fig. 7.5 Examples of lip images regarded to be unsuccessfully detected. It can be seen that the images are of varying dimensionality, with issues such as identifying the wrong area of an image as the ROI, tracking only part of the lip-region, or poor quality information due to blurring and head motion

Fig. 7.6 Examples of lip images where no ROI was identified and cropping was not successful. It can be seen that this is due to the speaker turning their head or obscuring their face

result (one where the fuzzy manual score matched), and 7.85 % produced what was considered to be an incorrect result, as shown in Table 7.1. Taking into account that 10 of the 20 sentences consisted of active conversation, this was a considered to be a good overall result.

To analyse the results in more detail, a comparison of the number of frames assigned each score is shown in Table 7.2. The difference in estimated values between the manual and the fuzzy approach is shown in Table 7.2. This table shows that 3.98 % of frames were incorrectly categorised as being good values (i.e. the difference between the ground truth and automatic values), 5.5 % were incorrectly estimated to identify no ROI, and 44.8 % were estimated to incorrectly be estimated as having a value of 2 (i.e. an incorrect/blurry/partial region). This was unsurprising as the difference between good and poor values could sometimes be very small, and indicates that the detector may have limitations with regard to precise identification of incorrect but partial regions.

Table 7.1 Overall performance of visual quality fuzzy input variable compared to manual scoring, considering each frame of all 20 speech sentences

	Number of frames	Percentage (%)
Correct	36,836	92.15
Incorrect	3139	07.85
Total	39,975	100

Table 7.2 Error between estimated visual fuzzy input and manual value for each frame of all 20 speech sentences

Estimated value	Manual est.	Fuzzy est.	Difference	Difference percentage (%)
1	36,334	37,779	1445	3.977
2	3168	1749	1419	44.79
3	473	447	26	5.497

To analyse the precise results, it was considered to be of interest to compare the error in individual sentences in order to identify if differences between the fuzzy estimation and the manual evaluation were evenly split, or were concentrated in specific sentences. Each of the 10 sentences in each conversation subset was evaluated to compare the difference in results. For reasons of space, the results of the reading task are reported fully in [5], but they show that the percentage of matching fuzzy and ground truth values predicted is above 94 % in all cases, with only a very small number of results where the fuzzy estimation does not match the manual evaluation. Table 7.3 is of particular interest and shows the match between the fuzzy estimation and the manual evaluation for the 10 sentences chosen for the conversation task.

Table 7.3 shows that the variation between individual sentences is much higher, which is to be expected considering the issues the tracker faces with conversational speech. There is particular error concentrated in one sentence, with 66.18 % of frames showing a difference between the manual and fuzzy estimation. An inspection of this specific cropped image sequence identified that the reason for this was the performance of the tracker. While the tracker initially identifies a correct ROI, due to the specific features of this face, a large number of frames are considered to be partial

Table 7.3 Comparison of assigned values for 10 sentence conversation dataset, showing difference in estimated value for manual inspection and fuzzy logic variable

Sentence	No. correct	Perc. correct (%)	No. incorrect	Perc. incorrect (%)	Total frames
1	1836	91.85	163	8.15	1999
2	1432	71.64	567	28.36	1999
3	1999	100	0	0	1999
4	1947	97.40	52	2.60	1999
5	1840	92.05	159	7.95	1999
6	1930	96.55	69	3.45	1999
7	676	33.82	1323	66.18	1999
8	1689	84.49	310	15.51	1999
9	1978	98.95	21	1.05	1999
10	1996	99.85	3	0.15	1999

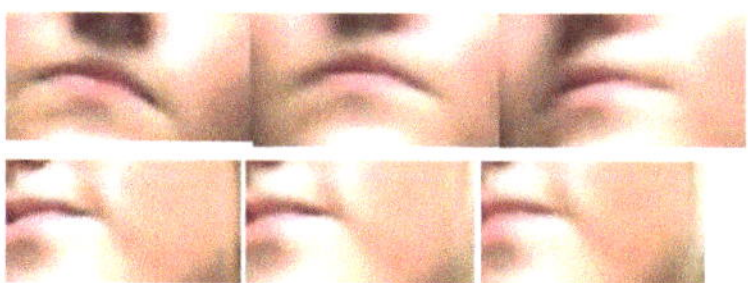

Fig. 7.7 Examples of lip tracker extracting an incorrect image for a sequence of frames. These frames were consecutive frames from a single sentence and show that while a manual investigation may identify this as a partial result, the fuzzy input may be more nuanced, due to most of the mouth being present

and only show a percentage of the mouth. While a manual inspection resulted in these being classified as partial results, the majority of the mouth was shown in these frames, as shown in Fig. 7.7, and so the difference was relatively small, resulting in the fuzzy value assigning these a score that was within the range of being considered good quality data. This indicates the difficulties with giving a precise score of 1, 2, or 3, and is a justification for using a fuzzy variable rather than a crisp set.

In summary, the visual input fuzzy variable was considered to be very accurate, with the majority of frames being correctly classified. It can be seen that the majority of errors were found in one specific sentence in the test-set. An inspection of the data demonstrated that this could be identified as due to potential ambiguity over the definition of the visual data, thus justifying the use of fuzzy logic rather than crisp sets. A comprehensive summary of these results can be found in [5].

7.4.2 Previous Frame Fuzzy Input Variable

7.4.2.1 Problem Description

The aim of the previous frame variable was to prevent rapid switching from frame-to-frame when the input data could theoretically be processed by more than one processing option and there were very small differences from frame-to-frame, meaning that a small change in environmental conditions may result in rapid changes in processing decision from frame-to-frame. Although it was decided to make use of the previous single output decision, it was possible that using a moving average of the previous outputs could be more effective in reducing switching than using a single value, and so this section investigates the effect of making use of the single previous output and compares this to using a mean of the previous 3, 5, and 10 previous output decisions. In addition, the aim of using this input variable was to reduce oscillation, and so the effectiveness of using this variable is evaluated by comparing the output from the system when the rules relating to this input variable were enabled to the equivalent output when the rules were disabled (and so ignoring the previous frame input variable entirely).

7.4.2.2 Experiment Setup

A small dataset of 3 sentences was used for evaluation. Broadband machine noise was added to these sentences using the simulated room environment at varying SNR levels to produce 18 noisy speech sentences with a range of audio and visual fuzzy input variables. In addition to this 3 sentences recorded in a noisy environment were also used, producing a total of 21 sentences. This input was then evaluated using the fuzzy logic system, and the output decision for each frame was recorded. In addition, the fuzzy rules pertaining to the previous frame were disabled, and the 21 sentences were evaluated again, and the decision (this time effectively only using two input variables) for each frame was also recorded.

The 21 sentences were evaluated, using the single previous output decision, the mean of the value for the previous 3 outputs, the mean of the previous 5 outputs, and the mean of the previous 10 outputs as the input variable. The resulting output processing decision from the fuzzy logic system was then compared to the decision from the previous frame to calculate the difference between frames. As the system is fuzzy, it is possible for the output decision to vary very slightly from frame-to-frame, without the difference being large enough to affect the processing decision (i.e. no processing, audio-only, or audiovisual), and so it was felt of more relevance to focus on frames where there was a difference in output decision from the previous frame greater than ± 1.

7.4.2.3 Summary of Results

Initially, an investigation was carried out into whether using the single previous output decision, or whether a moving average of the previous 3, 5, or 10 output decision values would result in the greatest reduction in difference between frames. The use of the previous fuzzy output decision as an input into the subsequent frame is intended to reduce rapid frame-by-frame fluctuations. Table 7.4 shows the mean values of the number of frames where a difference greater than or equal to ± 1 is found from the previous frame, showing the total number of frames with a difference and the percentage of the total frames, for the four different previous input variables, and takes account of when the previous frame rule is both enabled and disabled in the system. A full breakdown of results for individual test sentences is published in [5].

Table 7.4 Number and percentage of frames with difference in fuzzy output decision compared to previous frame, showing mean difference of all sentences (41,980 frames), evaluated with previous frame rule enabled and disabled in fuzzy-system

Fuzzy	Prev. frame		Mean of 3 frames		Mean of 5 frames		Mean of 10 frames	
	No. diff.	Perc. diff. (%)	No. diff.	Perc. diff. (%)	No. diff.	Perc. diff. (%)	No. diff.	Perc. diff. (%)
Enabled	1132	2.697	1191	2.837	1208	2.878	1210	2.882
Disabled	2460	5.856	2460	5.856	2460	5.856	2460	5.856

Table 7.4 shows that increasing the number of previous decisions used as part of the mean input variable results in a very small increase in difference. When only the single previous output decision is used as the input variable, 2.7 % of the frames show a change in decision. Using a mean of the 3 previous decisions results in a change of 2.8 %, increasing to 2.9 % when a mean of 5 previous decisions, and then finally 2.9 % when a mean of the 10 previous decisions is used. Overall, the difference between frames when using an increased number of previous decisions as part of the input mean variable was considered to be so small that it had no particularly noticeable difference. Therefore, it was felt that it was suitable to use only the previous decision as an input variable into the fuzzy logic system.

The second aspect of this evaluation concerned the impact that this fuzzy input variable had on reducing the oscillation from frame-to-frame. To investigate this, the test-set described above was evaluated with the system. The fuzzy logic system was adjusted to disable the rules concerning the previous input variable, in effect meaning that the system made use of only the audio and visual input variables. The mean results of this evaluation are also shown in Table 7.4. Firstly, because no previous frame rules are enabled, there is no change at all when a different number of previous decisions are part of the mean input variable. As expected, with the rules enabled, the frames with a recorded difference varies from 2.7 % to 2.88 %. With this input variable not used, 5.86 % of frames record a difference in output decision from the previous frame. Therefore, it can be concluded that the use of this input variable successfully limits processing decision variation from frame-to-frame, justifying the decision to make use of it as an input into the fuzzy logic speech filtering system.

7.5 Detailed System Evaluation

7.5.1 Problem Description

This section focuses on the evaluation of the multimodal fuzzy logic based speech enhancement framework. Due to the limitations discussed in Sect. 7.2, the custom corpus described in Sect. 7.3 was used.

Again, due to the limitations discussed previously, it was not possible to use a truly noisy corpus that was compatible with this preliminary system, and so noise was added to the system using the simulated room environment discussed in Chap. 4. However, the limitation with this approach is that the simulated room with clearly defined noise and speech sources represents an ideal scenario that an audio-only beamformer would be expected to process without difficulty. Because of this, and also the limitations with the audiovisual approach when presented with novel data (as discussed in Chap. 5), the results presented in this section should be interpreted with a degree of caution. There are some limited examples of the Lombard Effect, in that the speaker is taking part in an animated conversation and so adjusts their voice to take account of both parties talking, but this is not considered to be a major part of the chosen test-set.

7.5.2 Experiment Setup

The system was evaluated by making use of 10 sentences selected from the newly recorded audiovisual corpus described in Sect. 7.3. To provide a range of data, 5 of the 10 sentences were taken from the reading task, and 5 from the conversation task. Each sentence was split into frames, producing approximately 1999 frames per sentence. Noise was added to speech with the simulated room environment at a range of SNR levels, from −40 to +10 dB. Two different noises are used. Firstly, broadband noise is added to the speech. This consisted of a recording of a washing machine, as shown in Fig. 7.8. It can be seen in Fig. 7.8 that the amplitude of the signal varies over time, with a gradual decrease in amplitude throughout the recording segment. This noise was used for the subjective listening tests in Sect. 7.5.3.

To compare the results of using broadband noise with using a different noise source, an inconsistent clapping noise is also used, as shown in Fig. 7.9. A number of different versions were compared of each speech sentence. For the objective tests, the audiovisual approach presented in Chap. 4 was used, along with an audio-only beamforming approach to serve as a comparison. In addition to this, the spectral subtraction approach used in Chap. 5 and the unfiltered noisy signal were also used. These results were compared with results from the fuzzy logic based system.

As the speech sentences were considerably longer than the original speech sentences used for evaluation in previous chapters (20 s rather than 3 s), to reduce listener fatigue, only three versions of each conversation snippet were evaluated, audiovisual, audio-only, and the fuzzy logic approach, and the number of 20 s conversation snippets was reduced from 10 to 5. These were then tested with volunteers to produce suitable MOS. In addition to the audio output, the fuzzy switching is also evaluated. To do this, individual sentences were inspected to assess the effect of various factors, such as adjusting the SNR, the effect of poor visual information, and the difference between individual sentences. This is evaluated by a visual inspection and

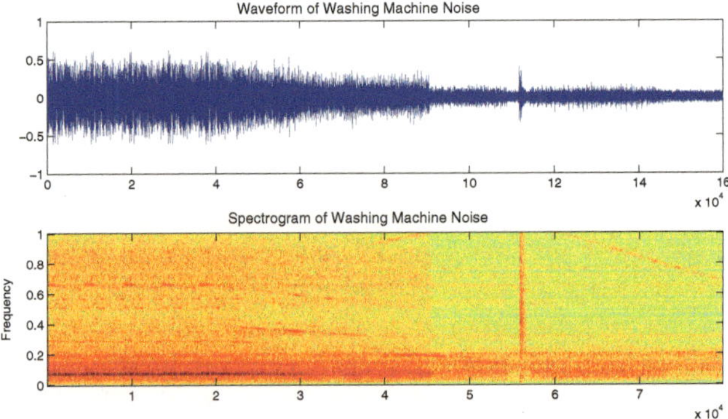

Fig. 7.8 Waveform (*top*), and spectrogram (*bottom*) of broadband washing machine noise used for objective and subjective tests

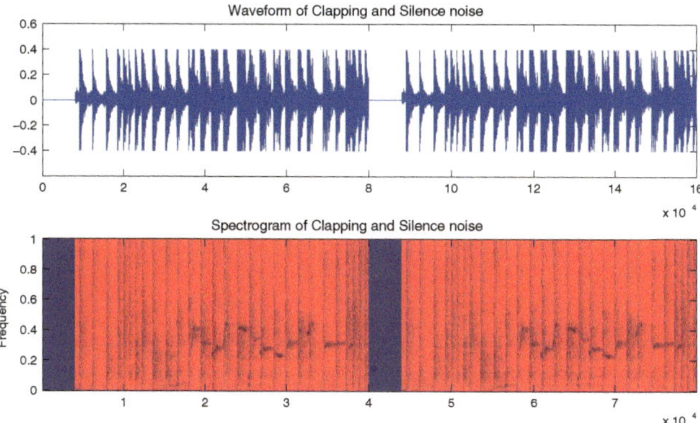

Fig. 7.9 Waveform (*top*), and spectrogram (*bottom*) of clapping noise used for objective tests

comparison. The fuzzy output decision on a frame-by-frame basis is then inspected. This is discussed in Sect. 7.5.4.

7.5.3 Subjective Testing with Broadband Noise

Listening tests were conducted in a manner similar to those discussed in Chap. 5. 10 volunteers took part in listening tests in a quiet room, using noise cancelling headphones. All of the volunteers spoke English as a first language, and none reported any abnormalities with their hearing. There were 6 male subjects and 4 female subjects, with an age range between 21 and 37. Listeners were played sentences randomly from the test-set, and were asked to score each between 0 and 5 based on the same criteria as in Chap. 5. They were asked to score speech signal distortion, noise intrusiveness level, and overall quality. MOS results are shown in Table 7.5. As there is a similar trend in results, only the overall scores will be examined in more depth here, with full data discussed in [5].

It can be seen that the audiovisual approach is consistently identified to have the worst output scores, and the audio-only technique returns the best results. The

Table 7.5 MOS table for overall quality for speech with washing machine noise added, for audiovisual speech, audio-only beamforming, and fuzzy-based processing

Level (dB)	Avis	Beamforming	Fuzzy
−40	1.392	3.400	1.196
−30	1.650	4.130	1.604
−20	1.774	4.374	1.988
−10	1.950	4.364	3.650
0	1.436	4.370	3.986
+10	0.872	4.414	3.787

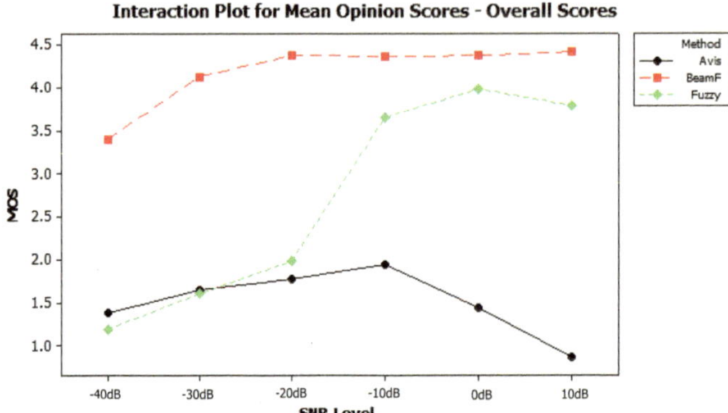

Fig. 7.10 Interaction plot for overall MOS at varying SNR levels, showing audiovisual speech (*black* and *circle markers*), audio-only beamforming (*red* with *square markers*), and fuzzy-based system (*green* with *diamond markers*)

Table 7.6 Selected results of Bonferroni Multiple Comparison, showing P-Value results for difference between Audio-only beamforming and Fuzzy Processed Speech for overall subjective scores

Level (dB)	Difference of means	SE of difference	T-Value	Adjusted P-Value
−40	−2.204	0.180	−12.23	0.000
−30	−2.526	0.180	−14.02	0.000
−20	−2.386	0.180	−13.24	0.000
−10	−0.714	0.180	−3.96	0.012
0	−0.384	0.180	−2.13	1.000
+10	−0.627	0.180	−3.479	0.081

fuzzy-based approach performs poorly at a very low SNR, but has an improved output at a higher SNR. A more detailed analysis is conducted on the results, using Bonferroni multiple comparison. Figure 7.10 shows the interaction plot for overall MOS, and the difference of means between audio-only and fuzzy is given in Table 7.6, and between audiovisual and fuzzy in Table 7.7.

It can be seen that at a lower SNR, the audiovisual and fuzzy-based scores are very similar, with no significant difference. This signifies that there was a far greater preference by listeners for the sentences processed with audio-only beamforming. When the SNR is increased, the fuzzy-based approach produced an improved score, with a similar output to the audio-only approach, with the results of Bonferroni multiple comparison showing that at SNR levels of −10, 0, and +10 dB, the overall and speech distortion scores were not significantly different ($p > 0.05$). This indicates that listeners found these sentences to be very similar in terms of overall results, and also when considering the speech in isolation.

Overall, the results confirm that the fuzzy-based system performs as expected. At lower SNR levels (−40 to −20 dB), the MOS is very similar to the audiovisual MOS, with small but not significant differences, as shown by the results of a comparison of means. At SNR levels of −10 and 0 dB, the audio-only and fuzzy-based results are very similar, suggesting that audio-only processing is used more often. At a high

Table 7.7 Selected results of Bonferroni Multiple Comparison, showing P-Value results for difference between Audiovisual Filtering and Fuzzy Processed Speech for overall subjective scores

Level (dB)	Difference of means	SE of difference	T-Value	Adjusted P-Value
−40	−0.196	0.180	−1.088	1.000
−30	−0.046	0.180	−0.255	1.000
−20	0.214	0.180	1.188	1.000
−10	1.700	0.180	9.434	0.000
0	2.550	0.180	14.151	0.000
+10	2.915	0.180	16.18	0.000

SNR, the significantly different noise intrusiveness score between audio-only and fuzzy-based scores shows that the system is making use of some unfiltered data. The results also show that audiovisual is the worst performing technique, and the audio-only approach far outperforms this method. However, these results should be interpreted with a degree of caution. As discussed in this chapter, the audio-only beamforming was expected to perform well, as the simulated room environment is designed specifically to demonstrate the performance of this technique.

Additional experiments, reported in full detail in [5], but summarised here for space reasons, make use of a noise that includes transient silence and clapping, as this was not a noise designed for the beamformer to process easily. The results showed that as expected, the audio-only beamformer returned the same score at all SNR levels, as no recognisable speech was returned. The audiovisual score was also very low, but listening to the output confirmed that an audio signal could be heard, hence the higher score. The results reported in that despite the lack of output signal, the difference between the fuzzy output and the audio-only output is only significant at a SNR of −40, and 0 dB (where $p < 0.05$). The difference between the audiovisual and fuzzy output scores was not significant at any SNR level.

The results demonstrated that the audio-only beamforming results presented in the previous sections should be interpreted with a degree of caution as fuzzy logic based results presented in this section are very dependent on the techniques used for processing speech. Although previous sections reported that the audio-only approach produced clearly better results, this was when the noise was one which the beamformer was capable of processing. Likewise, the audiovisual results were shown to be limited due to the system not being trained with data similar to that used for testing. As discussed, full results are available at [5].

7.5.4 Detailed Fuzzy Switching Performance

As discussed previously in this chapter, it can be seen that the fuzzy logic output varies depending on factors such as the SNR level and the previous output decision value, and the results of subjective and objective tests show that the output mean scores are often similar, but not identical to either the audio-only output scores or the audiovisual scores. However, as a range of sentences (with different associated visual quality fuzzy values), noises, and SNR levels were tested, it was felt suitable to examine the performance of the fuzzy switching approach in detail, again, for reasons of space, more examples are reported in [5].

7.5.4.1 Fuzzy Switching with Varying Noise Type

Firstly, the difference between sentences mixed with the two different noises used in this chapter is examined. To do this, two sentences are compared, with different noise added. The fuzzy output decision from frame-to-frame of a sentence with transient noise is compared to the frame-by-frame output decision of the same sentence, except with the machine noise added at the same SNR. Noise was added at a SNR of −20 dB to the sentence, and the output is shown in Fig. 7.11. In order to ensure that good quality visual information was available at all times, an example of a sentence from the reading task was chosen.

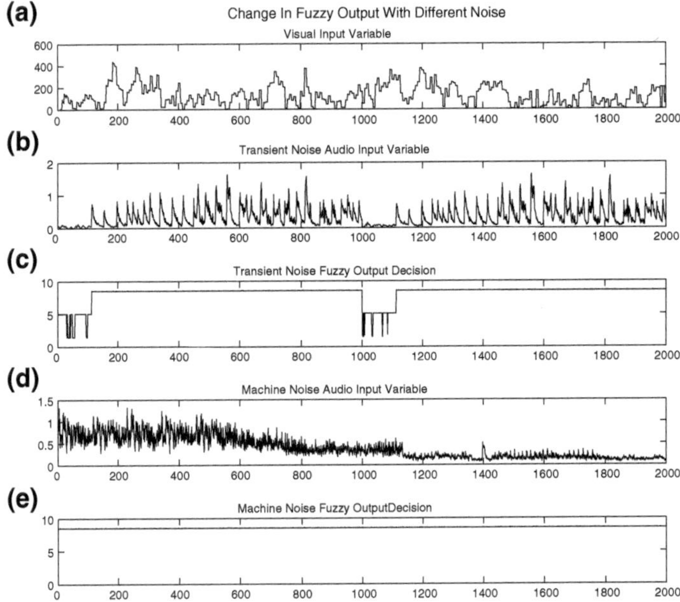

Fig. 7.11 Comparison of fuzzy logic output decision depending on noise type at SNR of −20 dB. **a** shows the input visual information. It can be seen that all values are below 600, therefore every frame is considered to be good quality. As the visual information is unchanged, then this is the same for both transient and machine noise speech mixtures. **b** shows the transient mixture fuzzy input variable. **c** shows the associated transient noise mixture output processing decision. **d** shows the machine noise mixture fuzzy input variable. **e** shows the machine noise mixture output processing decision

Figure 7.11 shows the difference in the fuzzy output decision, depending on the input noise variable. As the visual information, SNR, and sentence content was the same for both values, the only difference was the noise type. (c) Shows the fuzzy output decision, based on the visual input variable in (a), and the audio input variable in (b). With the transient noise, it can be seen in (c) that there are two large quiet periods, which is to be expected when it is considered that this noise consists of handclaps and silences (see Fig. 7.9). In these periods, either the unfiltered or audio-only options are chosen, otherwise, the audiovisual output is chosen as expected. (e) Shows the fuzzy output decision, based on the visual information in variable (a), and the audio input in variable (d). It can be seen that the fuzzy decision is different, as the noise input variable is different. The machine noise is a broadband noise, and so there is more noise present. As the SNR consistently remains very low, the audiovisual output is chosen in all frames. In summary, it can be seen that the fuzzy output decision varies based purely on the noise type. Additional examples can be found at [5].

7.5.4.2 Fuzzy Switching with Varying Visual Information

The previous examples considered a sentence with good quality visual information available at all times, but it was also considered to be of interest to observe the effect that varying the quality of visual information had on the fuzzy decision. As discussed in Chap. 6, if the audio input level was considered to be high, then the fuzzy logic system would use audiovisual processing, but only if the visual information was considered to be of good quality (i.e. the visual input fuzzy variable was low). To test this, a number of different sentences are compared, and the fuzzy outputs compared. In all sentences, machine noise is mixed with the speech signal at a SNR of −30 dB to ensure consistency, and provide a noise where audiovisual processing would be expected to be chosen for all frames if good quality visual information is available. Figure 7.12 show an example of this, with further examples available at [5].

In Fig. 7.12, (a) represents the visual input variable, (b) represents the audio input variable, and (c) shows the fuzzy output decision. It can be seen that despite the noise type and SNR justifying audiovisual processing, the visual input variable (which was shown to be accurate in Sect. 7.4) varies, and so the system only uses audiovisual processing when it is considered to be suitable. This demonstrates that the system only uses audiovisual information when it is considered to be appropriate, and adapts to different sentences.

7.5.4.3 Fuzzy Switching with Varying SNR Level

In addition to considering the effect of noise type and visual information, the effect of mixing the speech and noise sources at varying SNR levels is of interest. For this example, one sentence was chosen, with a small number of frames with poor quality visual information, and the noise source was the broadband machine noise.

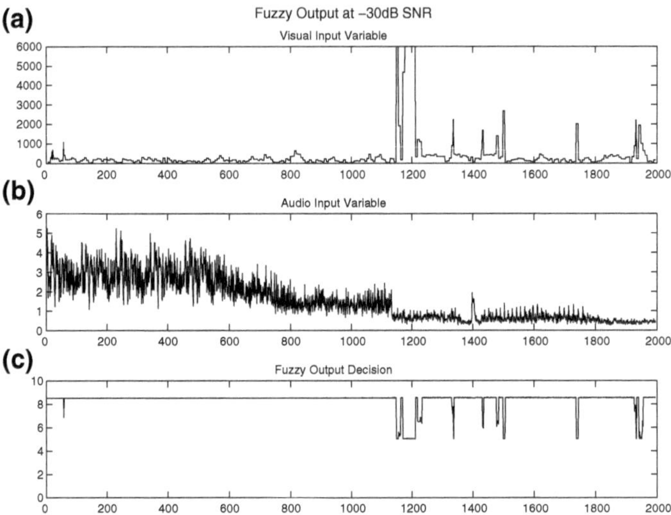

Fig. 7.12 Fuzzy logic output decision depending on quality of visual information, for sentence with several frames considered to be of poor quality. **a** shows the input visual variable. It can be seen that there are a number of frames where there is considered to be poor visual input. **b** shows the audio input variable, with machine noise added to speech at an SNR of −30 dB. **c** shows the fuzzy output processing decision

The sentences were then mixed at different SNR levels. In Fig. 7.13, (a) represents the mixed audio waveform, and (b) the associated fuzzy input variable. (c) Shows the visual input variable and (d) shows the fuzzy processing decision output.

It can be seen that initially, the audiovisual processing option is chosen where appropriate. Later in this sentence though, when there is considered to be lower quality visual information available, the system chooses audio-only processing. The decision does not quickly change back to audiovisual processing, but continues to choose audio-only processing for number of frames. This is because of the increased SNR, demonstrating that the fuzzy logic system adapts to different noise inputs. This adaptability is also shown in Fig. 7.14.

Firstly, in Fig. 7.14, it can be seen in (a) that the speech is more visible in the waveform, which is a reflection on the increased SNR level. It can be seen in (d) that as the input level variable decreases, the fuzzy logic system chooses the audio-only option for much of the second part of the sentence, which is very different from previous examples of the same sentence with the same noise but a lower SNR.

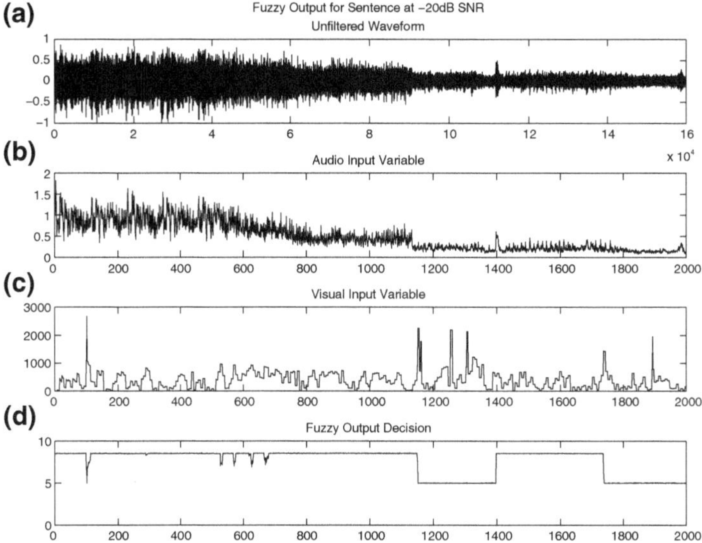

Fig. 7.13 Fuzzy logic output decision depending on SNR level. **a** shows the input audio waveform, with speech and noise mixed at a SNR of −20 dB. **b** shows the audio input variable. **c** shows the visual input variable, with a small number of frames considered to be of low quality. Finally, **d** shows the fuzzy output decision

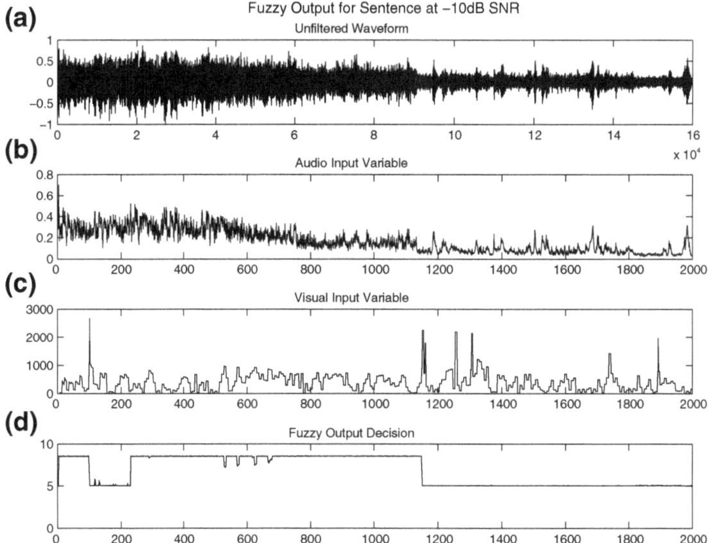

Fig. 7.14 Fuzzy logic output decision depending on SNR level. **a** shows the input audio waveform, with speech and noise mixed at a SNR of −10 dB. **b** shows the audio input variable. **c** shows the visual input variable, with a small number of frames considered to be of low quality. Finally, **d** shows the fuzzy output decision

7.6 Discussion of Results

Firstly, the evaluation performed in this chapter confirmed that the fuzzy-based system performs as expected. The system switches between processing options when considered appropriate. However, there are further improvements that could be made to this approach. There are occasions when data that was manually identified as being of poor quality, was classified as being correct by the detector. To improve the accuracy of the fuzzy input variable, it could be possible to create an improved input variable using a machine learning technique. Section 7.4 also discussed the use of the previous fuzzy controller output value as an input into the system for the subsequent frame, and showed that although using a floating mean smoothed the input variable on a frame-to-frame basis, it made very little difference to the fuzzy output value, justifying the use of a single frame.

However, there are a number of ways in which these inputs could be improved. As discussed above, a model could be trained to accurately identify the quality of an image. Also, in addition to the relatively basic audio power input, additional detectors such as a VAD could be used to positively identify the presence or absence of speech. This would serve as an additional input into the fuzzy-based system (and so would require the writing of additional rules), as used in some current commercial hearing-aids. This could also include specific front-back or wind detectors, to add versatility to the system.

With regard to the audio output of the system, it can be seen from the evaluation that the results are of limited value. This was discussed in Chap. 5, where the result was found to be significantly worse when used with data not similar to that which the system had not previously been trained with. Therefore, poor results were expected with novel data. Accordingly, although the fuzzy-based approach performed as expected, the limitations identified with the speech processing techniques also show that the system is currently only suitable for testing in specialised environments, and needs further development before being suitable for more general purpose use.

References

1. I. Almajai, B. Milner, Effective visually-derived Wiener filtering for audio-visual speech processing, in *Proceedings of Interspeech, Brighton, UK* (2009)
2. S. Naqvi, M. Yu, J. Chambers, A multimodal approach to blind source separation of moving sources. IEEE J. Sel. Top. Signal Process. **4**(5), 895–910 (2010)
3. I. Almajai, B. Milner, Maximising audio-visual speech correlation, in *Proceedings of AVSP* (2007)
4. B. Rivet, L. Girin, C. Jutten, Mixing audiovisual speech processing and blind source separation for the extraction of speech signals from convolutive mixtures. IEEE Trans. Audio Speech Lang. Process. **15**(1), 96–108 (2007)
5. A. Abel, A. Hussain, Cognitively inspired fuzzy based audiovisual speech filtering, Technical report CSM-198, University of Stirling, Computing Science and Mathematics, March 2014

Chapter 8
Potential Future Research Directions

Abstract The speech enhancement research presented here was motivated by several factors. Firstly, the development in recent years of audio-only hearing aids that utilise sophisticated decision rules to determine the appropriate level of speech processing served as an inspiration. A second motivational factor was the exploitation of the established cognitive relationship between audio and visual elements of speech to produce multimodal speech filtering systems. Another motivation was the desire to utilise audiovisual speech filtering to extend the concept of audio-only speech processing to become multimodal, from the perspective of potential application to hearing aids. Based on these motivations, the goal of the work presented in this book was primarily to develop a flexible two-stage multimodal speech enhancement system, working towards the development of a cognitively inspired fuzzy logic based speech enhancement framework that is autonomous, adaptive, and context aware. The novel proof of concept framework presented in this book makes use of audio-only beamforming, visually derived Wiener filtering, state-of-the-art lip tracking with Viola-Jones ROI detection, and a fuzzy logic controller, to present a novel speech enhancement framework. This chapter presents some conclusions and future work.

Keywords Multimodal speech filtering · Fuzzy logic · Conclusions · Potential improvements · Future work

8.1 Improvement of Individual Speech Processing Components

As discussed in Chap. 5, limitations have been identified with some of the individual speech processing components presented in this system which could be improved. One significant example of this is the Wiener filtering approach. The current implementation is fairly basic, utilising GMM-GMR to provide an estimation of the noise free speech signal in the filterbank domain and interpolating this. A single GMM is also used for speech estimation. However, this has limitations due to the relative simplicity of its implementation. The GMM-GMR approach was originally devised

A. Abel and A. Hussain, *Cognitively Inspired Audiovisual Speech Filtering*,
SpringerBriefs in Cognitive Computation, DOI 10.1007/978-3-319-13509-0_8

by [1] to calculate efficient robot arm movement. Furthermore, the speech modelling technique used does not make use of some of the most recent developments in speech enhancement, which may improve results. So for example, [2] make use of phoneme specific GMMs that attempt to identify the phoneme spoken, and then apply a specific GMM to this portion of speech.

Other state-of-the-art beamforming techniques could be investigated to be considered for integration within this framework, and alternatives to using GMMs, such as reservoir computing [3], an area which has been recently applied to the signal processing domain for tasks such as multimodal laughter detection and music classification [4, 5] can also be considered, to improve on the visually derived filtering approach and improve results.

8.2 Extension of Overall Speech Filtering Framework

One outcome of the work is the initial development of a novel, scalable, speech processing framework that extends from feature extraction to speech filtering, with the use of a fuzzy logic controller. However, there is still much potential for extension of this framework. In addition to future work to upgrade the existing components of the system and investigate new speech enhancement techniques, it is also possible to add additional components to the framework. Some examples include the possibility of adding additional inputs such as spike trains [6, 7] to potentially improve the filtering process. Other speech processing research (for example, by [8] and also the author in [9]) has found that asynchrony can also result in an improved audiovisual speech relationship, and this could be exploited in future work.

Another way the framework can be extended is to include a number of more sophisticated input detectors, such as wind and front-back detectors, as discussed in Chap. 3. Work by [10] has resulted in a speech enhancement system that uses a VAD to identify areas of speech and non speech in the input signal. This additional detector has precedent for being used in the literature, and may improve the fuzzy logic aspect of this system greatly. If the system was to be extended to successfully process real world data, then some of the other detectors discussed by [11] such as wind detectors and front back detectors could also be integrated into the system, all of which would add sophistication and feasibility to the framework presented here.

8.3 Further Development of Fuzzy Logic Based Switching Controller

The fuzzy logic controller presented as part of this novel speech enhancement framework in Chap. 6 is a very basic implementation, demonstrating that this framework could be developed further. Although it has been demonstrated that the fuzzy-based system is capable of responding to environmental conditions as expected, the results of running tests with real data have to be treated with some caution, due to the limitations of the test environment, the preliminary nature of the system (in that it is not implemented in real time), and the limitations with the filtering techniques identified in Chap. 5. Although tests have been carried out using more challenging data, in order to test the system further, hardware based tests using multiple microphones and more real data are needed.

Additionally, the range and quality of input variables and fuzzy sets could be improved. As stated in Chap. 6, the three variables used, audio frame power, visual DCT detail level, and previous frame selection, represent fairly simple detectors to use as fuzzy inputs. Although these are sufficient for demonstrating the novel framework presented in this work, an extension of this would naturally investigate the use of the detectors mentioned earlier, such as modulation and wind detectors, as inputs to the fuzzy switching system. These could then be used to develop the rules further. Although the current rules are adequate for demonstrating proof of concept, there are potential areas of improvement such as tweaking the weighting of the rules to give priority as suitable, rewriting the rules to cope with potential new inputs, and considering other aspects such as engaging/adaptation/attack time when it comes to selection of the processing option. As discussed in Chap. 7, additional refinement could also be carried out with regard to the visual input variable. Although the initial implementation was found to function well, it could be improved further by using a machine learning technique such as a trained HMM or ANNs to classify input lip images.

It is also important that any future work carries out further testing of the refined framework. As stated previously, the current proof of concept framework has only undergone limited evaluation with it being concluded that further refinement is needed. Listener comfort is of particular importance. The system is designed with the considerations of users with hearing loss in mind, and is designed to automatically switch between processing options as needed, but it is important that this is done in a manner that does not cause irritation to the listener. Although the fuzzy inputs were shown to minimise frame to frame oscillation, this could be investigated and evaluated further in future work.

8.4 Practical Implementation of System

The system is currently purely implemented through software and simulations. MATLAB has been used for development, and testing has been carried out using a pre-recorded corpus, mixed with noise using a simulated room. Future development of this system would be to extend this initial software implementation (and the proposed refinements discussed previously in this section) and work towards the development of an initial hardware prototype. This would implement the improved fuzzy logic based speech enhancement framework physically, and would be expected to function with live data and real world noise, rather than simply with pre-recorded corpora. An example of a potential implementation strategy would be to make use of FPGAs.

These are semiconductor devices that can be programmed after manufacturing and thus allow for rapid prototyping and debugging. For this reason, they are commonly used in initial hardware development of technology. The evaluation process could also be improved by hardware implementation, in that it would be possible to carry out listening tests in a truly noisy environment, taking full account of the Lombard Effect, room impulse responses, and data synchrony, providing a full evaluation of a real time system able to function with a wide range of challenging data.

References

1. S. Calinon, F. Guenter, A. Billard, On learning, representing, and generalizing a task in a humanoid robot. IEEE Trans. Syst. Man Cybern. Part B **37**(2), 286–298 (2007)
2. I. Almajai, B. Milner, Effective visually-derived Wiener filtering for audio-visual speech processing, in *Proceeding Interspeech*. (Brighton, UK, 2009)
3. W. Maass, T. Natschläger, H. Markram, Real-time computing without stable states: a new framework for neural computation based on perturbations. Neural Comput. **14**(11), 2531–2560 (2002)
4. S. Scherer, M. Glodek, F. Schwenker, N. Campbell, G. Palm, Spotting laughter in natural multiparty conversations: a comparison of automatic online and offline approaches using audiovisual data. ACM Trans. Interact. Intell. Syst. (TiiS) **2**(1), 4 (2012)
5. M. Newton, L. Smith, A neurally inspired musical instrument classification system based upon the sound onset. J. Acoust. Soc. Am. **6**, 4785–4798 (2012)
6. W. Maass, Networks of spiking neurons: the third generation of neural network models. Neural Netw. **10**(9), 1659–1671 (1997)
7. L.S. Smith, D.S. Fraser, Robust sound onset detection using leaky integrate-and-fire neurons with depressing synapses. IEEE Trans. Neural Netw. **15**(5), 1125–1134 (2004)
8. M. Sargin, Y. Yemez, E. Erzin, A. Tekalp, Audiovisual synchronization and fusion using canonical correlation analysis. IEEE Trans. Multimed. **9**(7), 1396–1403 (2007)
9. A. Abel, A. Hussain, Q. Nguyen, F. Ringeval, M. Chetouani, M. Milgram, Maximising audiovisual correlation with automatic lip tracking and vowel based segmentation, in *Proceeding of Biometric ID Management and Multimodal Communication: Joint COST 2101 and 2102 International Conference, BioID_MultiComm 2009, Madrid, Spain, 16–18 September 2009*, vol. 5707, (Springer, 2009) pp. 65–72
10. I. Almajai, B. Milner, Enhancing audio speech using visual speech features, in *Proceedings of the Interspeech* (Brighton, 2009)
11. K. Chung, Challenges and recent developments in hearing aids. Part i. Speech understanding in noise, microphone technologies and noise reduction algorithms. Trends Amplif. **8**(3), 83–124 (2004)

Index

A. Abel and A. Hussain, *Cognitively Inspired Audiovisual Speech Filtering*, SpringerBriefs in Cognitive Computation, DOI 10.1007/978-3-319-13509-0

L

M

N